Micromachining - New Trends and Applications

Edited by Zdravko Stanimirović and Ivanka Stanimirović

Published in London, United Kingdom

Micromachining - New Trends and Applications
http://dx.doi.org/10.5772/intechopen.114670
Edited by Zdravko Stanimirović and Ivanka Stanimirović

Contributors
Anup Dutt, Bagur Rudrappa Deepu, Belthangady Pavithra, Ivanka Stanimirović, Kunchinadka Narayana Bhat, Mangalore Manjunatha Nayak, Masaya Takasaki, Rakesh Kumar Bhardwaj, Somak Datta, Suresh Pratap, Takeshi Mizuno, Yuji Ishino, Zdravko Stanimirović

First published in London, United Kingdom, 2024 by IntechOpen
IntechOpen is the global imprint of INTECHOPEN LIMITED, registered in England and Wales, registration number: 11086078, 167-169 Great Portland Street, London, W1W 5PF, United Kingdom

British Library Cataloguing-in-Publication Data
A catalogue record for this book is available from the British Library

Additional hard and PDF copies can be obtained from orders@intechopen.com

Micromachining - New Trends and Applications
Edited by Zdravko Stanimirović and Ivanka Stanimirović
p. cm.

This title is part of the Industrial Engineering and Management Book Series, Volume 3
Topic: Production Engineering
Series Editor: Fausto Pedro Garcia Marquez
Topic Editor: Orhan Korhan

Print ISBN 978-0-85014-270-9
Online ISBN 978-0-85014-271-6
eBook (PDF) ISBN 978-0-85014-272-3
ISSN 3029-0511

For EU product safety concerns:
IN TECH d.o.o., Prolaz Marije Krucifikse Kozulić 3, 51000 Rijeka, Croatia,
info@intechopen.com or visit our website at intechopen.com.

IntechOpen Book Series

Industrial Engineering and Management

Volume 3

Aims and Scope of the Series

Industrial Engineering and Management (IEM) is a discipline that focuses on optimizing complex processes and systems within various industries. It involves the integration of engineering, business, economics, mathematics, and behavioral sciences to improve efficiency, productivity, quality, and overall performance in organizations. Key aspects of Industrial Engineering and Management include: Process Optimization; System Analysis and Design; Quality Control and Management; Supply Chain Management; Operations Management; Human Factors and Ergonomics; Project Management; Cost Analysis and Financial Management; Decision Analysis.

Overall, Industrial Engineering and Management aims to optimize resources, improve processes, enhance productivity, and ensure the effective and efficient utilization of all elements involved in the production or delivery of goods and services. It is crucial in today's competitive business environment for organizations to stay efficient and competitive.

Production Engineering and Operational Excellence are fields of study and practices that focus on optimizing and improving the manufacturing and production processes within an organization. It combines principles from engineering, management, and operational strategies to enhance productivity, efficiency, quality, safety, and sustainability in the production of goods and services.

Here are the key components of Production Engineering and Operational Excellence: Process Optimization; Operational Excellence; Manufacturing Systems Design; Quality Management; Supply Chain Optimization; Production Planning and Scheduling; Automation and Technology Integration; Health, Safety, and Environmental Management; Cost Management; Performance Measurement and Key Performance Indicators (KPIs); Continuous Improvement and Innovation.

Production Engineering and Operational Excellence are crucial for organizations aiming to stay competitive in the global market by achieving high levels of efficiency, quality, and customer satisfaction while optimizing resources and minimizing waste. It is a multidisciplinary approach that encompasses engineering principles, management strategies, and the effective use of technology to drive operational success.

Meet the Series Editor

Fausto Pedro Garcia Marquez is a Full Professor at UCLM, Spain, with accreditation since 2013. He also holds the position of Honorary Senior Research Fellow at Birmingham University, UK, and serves as a Lecturer at the Postgraduate European Institute. In addition to these roles, Fausto has experience as a Senior Manager at Accenture from 2013 to 2014. He earned his European Ph.D. with the highest distinction. Throughout his career, Fausto has received numerous awards and honors. These include the Nominate Prize (2022), Gran Maestre (2022), Grand Prize (2021), Runner Prize (2020), and Advancement Prize (2018), as well as Runner (2015), Advancement (2013), and Silver (2012) by the International Society of Management Science and Engineering Management (ICMSEM). He was also the recipient of the First International Business Ideas Competition 2017 Award. Fausto's contributions extend to academic publishing, with over 242 papers to his name. Notably, his work has been recognized in journals like "Applied Energy" (Q1, IF 9.746, Best Paper 2020) and "Renewable Energy" (Q1, IF 8.001, Best Paper 2014). His affiliations include the editorial and authorship roles in more than 50 books, with publications through respected publishers such as Elsevier, Springer, Pearson, Mc-GrawHill, IntechOpen, IGI, Marcombo, and AlfaOmega. He has authored over 100 international chapters and holds 6 patents. Fausto serves as the Editor of 5 International Journals and is a Committee Member for more than 70 International Conferences. His research portfolio encompasses being the Principal Investigator in 4 European Projects, 8 National Projects, and participating in over 150 projects involving universities and companies. His areas of expertise and research interests span Artificial Intelligence, Maintenance, Management, Renewable Energy, Transport, Advanced Analytics, and Data Science. Fausto is a recognized Expert in the European Union in AI4People (EISMD) and ESF. He also serves as the Director of www.ingeniumgroup.eu, holds the status of Senior Member at IEEE since 2021, and has been honored as an Honorary Member of the Research Council of the Indian Institute of Finance since 2021. Fausto is also the Committee Chair of The International Society for Management Science and Engineering Management (ISMSEM) since 2020.

Meet the Volume Editors

Zdravko Stanimirović, Ph.D., has been active in research and development work for the last 29 years. He received his MS and Ph.D. in Electrical Engineering from the Faculty of Electrical Engineering, University of Belgrade, Serbia, in 1999 and 2007, respectively. Dr. Stanimirović is currently an associate research professor at the "Vinča" Institute of Nuclear Sciences - National Institute of the Republic of Serbia, University of Belgrade. Over the years, he participated in several scientific projects funded by the Ministry of Science, Technological Development and Innovations of the Republic of Serbia. Dr. Stanimirović is the recipient of the IEEE Transactions on Components & Packaging Technologies Best Paper Award. His current research interests include material characterization measurements, MEMS, and micro/nanoscale sensors.

Ivanka Stanimirović, Ph.D., has been involved in research and development work for more than 25 years. Currently, she is an associate research professor at the "Vinča" Institute of Nuclear Sciences – National Institute of the Republic of Serbia, University of Belgrade, Serbia. Dr. Stanimirović earned her MS and Ph.D. in Electrical Engineering from the Faculty of Electrical Engineering, University of Belgrade, in 1999 and 2007, respectively. Over the years, she has worked on several scientific projects funded by the Ministry of Science, Technological Development and Innovations of the Republic of Serbia. She is the recipient of the IEEE Transactions on Components & Packaging Technologies Best Paper Award. Her current research interests include materials science, micro/nanoscale sensors, and reliability issues in MEMS.

Contents

Preface

Offering unparalleled precision and flexibility, micromachining has become a major asset in manufacturing technology. Because of the ability to reliably manufacture complex microscale features with very tight tolerances, micromachining has a strong impact on many different industries, encouraging efficiency and innovation. Precision and versatility in this field are greatly enhanced by the multitude of materials and techniques currently employed. At the same time, micromachining technology successfully intersects with environmental sustainability by introducing eco-friendly practices and reducing energy consumption and material waste. As a result, constantly increasing market demands in the miniaturization of products and systems strongly indicate micromachining's continuous progression in years to come.

To present their latest work in the field of micromachining, researchers from distant parts of the world have joined their efforts and contributed their ideas according to their interest and engagement. This book is the result of their hard work. You will have the opportunity to learn about materials, techniques, applications, challenges, and recent advancements in micromachining technology as well as about the state of the current micromachining market. You will have the opportunity to understand the concepts of micro-scale electronic component manufacturing. You will discover advancements in micromachining techniques of micro-electromechanical system (MEMS) piezoresistive pressure sensors introduced to minimize offset drift due to humidity and temperature. The book also presents the principles and classifications of force measuring systems with zero-compliance suspension. Finally, you will learn about triangular microcavity fabrication using micro-electrical discharge machining.

The editors would like to thank the authors for their contributions and efforts to present their work in a manner that allows both professionals and readers not involved in the field to understand the topic. We would also like to express special appreciation to the IntechOpen team for their dedicated work in making this book possible.

Dr. Zdravko Stanimirović and Dr. Ivanka Stanimirović
"Vinča" Institute of Nuclear Sciences,
National Institute of the Republic of Serbia,
University of Belgrade,
Belgrade, Republic of Serbia

Section 1

Micromachining Trends and Applications

Chapter 1

Introductory Chapter: Current Trends in Micromachining Technology

Ivanka Stanimirović and Zdravko Stanimirović

1. Introduction

Originating in the mid-1900s, micromachining became a cutting-edge method that is changing the way we think about small-scale component fabrication. Micromachining usually refers to a precise micrometer scale subtractive manufacturing processes that are essential for producing highly precise micro components, which are needed in the electronics, medical, aerospace and other growing industries. Fabrication capabilities are constantly being improved by the meticulous design of tools and processes that can handle dimensions and tolerances that traditional machining cannot handle. The industry's constant search for increasingly accurate and finer techniques is reflected in the progression from basic mechanical micromachining to more complex methods.

Through its special abilities and applications, micromachining sets itself apart from traditional machining. To meet the needs of a wide range of industries, micromachining can be used on a wider range of materials, such as polymers, metals and composites. Far beyond what is possible with conventional machining, micromachining produces features and tolerances on a micrometer scale. In these processes, much smaller and more precise micro tools are used, which frequently call for specific handling and setup. Precision that micromachining can provide is crucial for industries that require high levels of accuracy, such as semiconductor manufacturing and medical device fabrication.

This chapter provides a brief summary of the materials and techniques currently used in micromachining, as well as an overview of the main applications, challenges, and advancements in micromachining technology. It also provides a brief analysis of the current micromachining market.

2. Micromachining techniques and materials

The field's precision and versatility are greatly enhanced by the multitude of techniques [1–13] that are currently employed in micromachining, each well-suited to particular material and application. These techniques can be grouped into five major categories:

1. Mechanical micromachining [1–6]– a method used in precision engineering to create small, detailed features on workpieces. It is crucial for high-precision

industries like electronics and medicine because it allows realization of miniature parts with outstanding precision and surface qualities.

2. Thermal micromachining [7, 8]– a highly precise method for processing materials that are challenging to machine. Intense and focused energy source is used to modify or remove materials and create intricate details.

3. Chemical and electrochemical micromachining [9]– methods where chemical and electrochemical reactions are used for precise material removal and adequate surface finish.

4. Hybrid micromachining [10, 11]– a synergy of different machining techniques formed to improve the overall micromachining process.

5. Innovative techniques [12, 13]– techniques designed to meet specific requirements.

Each of these micromachining categories includes several key techniques presented in **Table 1**.

Micromachining techniques can handle fine micron scale features with extreme precision. Depending on the technique and material used, as well as on the component complexity, general tolerances are typically in the ± (0.001–0.005) mm range. Tolerances for high precision can be as low as ±0.0001 mm. Typical method-specific tolerances are presented in **Figure 1**. Micromachining tolerances [14] depend on multiple factors: material properties, tooling quality and precision, workpiece stability, machining environment, etc.

The precision and quality of the micromachined component are affected by how different materials react to micromachining techniques. Commonly used materials are metals (stainless steel, copper, titanium, etc.), ceramics (alumina, silicon carbide, etc.), polymers (ABS, Polycarbonate, etc.) and composites (Carbon Fiber Reinforced Plastics, etc.) [15–19]. Stainless steel, favored for its strength, durability and corrosion resistance, is one of the most popular materials used in micromachining [15]. Stainless steel is commonly used in medical applications for realization of various medical devices, implants and surgical instruments. Although highly-priced and difficult to machine, titanium is a valuable asset for micromachining [16]. It is light, heat and corrosion-resistant, biocompatible and inert and has a high strength-to-weight ratio. For these reasons, titanium can be used in a variety of applications ranging from medical to aerospace. Another metal suitable for precision machining, highly regarded for its natural corrosion resistance, thermal and electrical conductivity, durability and versatility, is copper. It is commonly used in electronics and cooling systems. Cost-efficient, nonconductive, nonmetal materials for micromachining are polymers [17]. Because of their inertness and adaptable properties, micromachined parts based on polymer materials are used in various industries ranging from medical to automotive. Micromachined alumina and silicon carbide parts are used in various industrial applications because of their thermal and chemical stability, dielectric strength, hardness, dimensional stability, ability to withstand harsh environments, etc. [18]. Micromachined ceramics allow fine feature formation without occurrence of heat-affected zones or microcracks while maintaining high cut quality. Carbon fiber-reinforced plastic (CFRP) materials are being recognized for their lightweight properties and high strength-to-weight ratio [19]. Micromachined CFRP parts are

Category: Mechanical micromachining		
Technique	**Description**	**Application**
Micro milling	CNC machines are being used for precise sculpting of complex shapes and geometries.	Micro molds, channels, and other intricate components.
Micro turning	Production of high surface quality and dimensionally accurate miniature cylindrical parts. Often combined with micro milling technique.	Small-scale components for medical, aerospace, and precision engineering applications.
Micro drilling	Creates precise microscale holes often using laser and ultrasonic technologies for advanced drilling.	Components for microfluidics, micro vias in PCBs.
Micro grinding	Creates components with exceptionally fine surface finishes using fine abrasive particles. Often used with other micromachining techniques.	Optical and other precision components.
Micro-USM*	Creates glass, hard metal or ceramic components using ultrasonic vibrations and abrasive slurry.	Semiconductor and MEMS components.
Category: Thermal micromachining		
Micro-EDM*	Shaping hard metals using electrical sparks.	Micro molds, dies, precision components with high aspect ratios.
Micro-LBM*	Technique that uses focused laser beams to form high aspect ratio microscale features.	Metal, plastic, ceramic components, PCB production.
Micro-EBM*	Technique that uses electron beams for precise material removal	Complex components for aerospace and medical applications.
Category: Chemical and electrochemical micromachining		
Chemical micromachining	Non-contact technique that uses controlled chemical reactions for etching micro-scale features.	Micro-scale glass, metal, silicon components.
Electrochemical micromachining	For high-precision micromachining free of the thermal stress anodic dissolution in an electrolytic solution is used in this technique.	Hard metal and alloy micro-components for medical, aerospace and automotive applications.
Category: Hybrid micromachining		
Combined Hybrid Micromachining	Technique that combines different micro machining processes, often mechanical and thermal processes.	Precision engineering components.
Assisted Hybrid Micro Machining	Technique that combines primary micro machining and an auxiliary process that temporarily modifies material properties for easier material removal.	Aerospace and medical components.
Category: Innovative techniques		
Micro-AJM*	Technique that uses high-velocity abrasive particles for material erosion	Glass and ceramic components.
Micro-FIB*	Technique that uses ion beams for accurate material deposition or milling.	Semiconductor manufacturing, MEMS fabrication.

**Abbreviations: Micro-USM—Micro Ultrasonic Machining, Micro-EDM—Micro Electrical Discharge Machining, Micro-LBM—Micro Laser Beam Machining, Micro-EBM—Micro Electron Beam Machining, Micro-AJM—Micro Abrasive Jet Machining, Micro-FIB—Micro Focused Ion Beam Machining.*

Table 1.
Micromachining techniques.

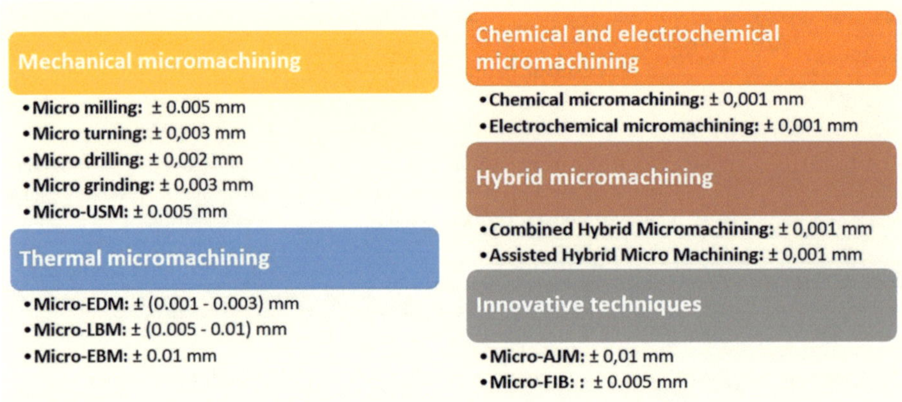

Figure 1.
Method-specific tolerances.

used in advanced engineering applications like electronics, aeronautics, automobiles, etc. Lately, micromachined glass has become of interest in the medical, aerospace, or microelectronic industries because micromachined glass parts can be manufactured with great accuracy while staying strong, are reusable and can be easily cleaned [19].

3. Primary micromachining applications

Contemporary micromachining techniques offer high precision and ability to work with diverse materials. They drive innovation and efficiency in a broad range of industries. Some of primary micromachining applications are in electronics, optics, telecommunications, medicine, MEMS, aerospace and automotive industries etc. (**Figure 2**). Micromachining is fundamental in microelectromechanical system (MEMS) manufacturing. MEMS microscale devices, actuators and sensors are being used in medical, electronics and automotive industries. In electronics, micromachining is essential in realization of miniature electronic devices such as microprocessors and microcontrollers. Electronic devices often require sub-micron range tolerances. Micromachined components for communication devices, such as micro-scale antennas and fiber-optic connectors, are essential in the field of telecommunications. Optical devices require precise geometries and excellent surface finishes. Performances of small-scale mirrors, lenses and other optical devices depend on high-precision micromachining techniques. Medical applications demand precise, complex, safe and effective micromachined implants and tools for surgical, diagnostic and other medical purposes. They require great precision, with tight tolerances, lower than ±0.002 mm. The aerospace industry requires miniature high-precision aerospace components of outstanding reliability and performance, e.g. fuel injectors, sensors, control mechanisms, etc. The automotive industry uses micromachining for creating small components for electric vehicles automotive sensors, fuel injection nozzles, etc. It can be concluded that micromachining's capacity to deal with diverse materials and produce intricate micro-scale geometries makes it a leader in the development and prototyping of new devices in a number of industries, including nanotechnology and new material creation.

Figure 2.
Primary micromachining applications.

4. Micromachining advantages and challenges

Micromachining, as a key method in advanced manufacturing, has several advantages:

- It is highly precise with tight tolerances down to the sub-micrometer range.
- Micromachining is able to process diverse materials, ranging from ceramics and metals to glass, polymers and composites.
- An outstanding feature is the capability to produce micro-scale complex components.
- Micromachining techniques can deal with diverse industrial demands.
- Micromachining reduces material waste and enhances cost-effectiveness and environmental sustainability.

However, micromachining also comes with a number of challenges:

- Micromachining requires specialized equipment and highly skilled staff resulting in high initial costs.
- Micromachining techniques are complex.
- Facilities have special requirements (e.g. ability to house large micromachining equipment, controlled environment, etc.).
- Tool maintenance and calibration are mandatory.

While micromachining entails a number of inherent challenges, these can be effectively managed with the appropriate approaches. It is important to ensure that available micromachining equipment can process materials and achieve tolerances needed for selected applications. The components' size and complexity must be taken into consideration when machining them. Machinery must be able to handle a range of microtools and work with CAD/CAM software to streamline processes. Feed rates and speed need to be carefully controlled to guarantee precision and reduce tool wear. Microtools are prone to rapid wear because of their small size and precision. This can be lessened by using premium materials and fine-tuning the machining parameters. Speed and efficiency in high-volume production environments must be carefully evaluated. It is important to choose energy-efficient equipment compatible with long-term growth strategies.

5. Micromachining market

Recent trends in product miniaturization in the electronics and semiconductor industries, where micromachining is essential for producing tiny parts, are driving the market. The micromachining market is expanding due to factors like faster data rates, greater compatibility with various wireless technologies, and longer battery lives of products, which is pushing IC manufacturers to rethink traditional production techniques. As a result, more electrochemical and laser machining processes are being used to produce micro components and provide precision machining. The application of Artificial Intelligence and the integration of the Internet of Things, Augmented Reality, Virtual Reality and 5G industries are also expected to benefit the market [20, 21]. Also, due to its widespread use in automotive industry, micromachining is becoming more and more in demand. It is used in fuel injection devices and electrical parts found in electrical vehicles.

Micromachining market share can be segmented on the basis of type, process, axis, end-use and regions [22–24]:

- By type, micromachining market can be divided into three segments: traditional, nontraditional and hybrid micromachining. The nontraditional micromachining encompasses laser, electrochemical (ECM), electro discharge (EDM) and laser micromachining. The nontraditional segment is being propelled by the precision, efficiency and wide applicability of the dominant laser micromachining. Ultrafast lasers are being developed especially for micromachining purposes stimulating the micromachining market expansion. The traditional segment owns its rapid growth to fast and adaptable material removal micro-milling process.

- By process, micromachining market can also be divided into two main segments: the dominant subtractive and additive micromachining. The subtractive segment owns its leading position in the cost and time-effective mass production, the ability to process various materials and produce large-scale components. The additive micromachining is also rapidly growing. Known as 3D printing for micromachining, additive micromachining is being used for versatile applications in MEMS sensors and industries such as medical, semiconductor, electronics, etc.

- By axes, micromachining market can be divided into three main segments: 3-axes, 4-axes and 5-axes micromachining (**Figure 3**). Until recently micromachining industry was led by 3-axes micromachining especially in advanced optics market. Nowadays, 5-axis micromachining replaces the casting process in realization of ultra-complex solid components mostly used in aerospace, automobile, medical and military equipment.

- By end-use, the micromachining market is oriented towards numerous industries: semiconductors, electronics, telecommunications, energetics, medical, automotive, military, aerospace, polymers, etc. (**Figure 4**). Electronic

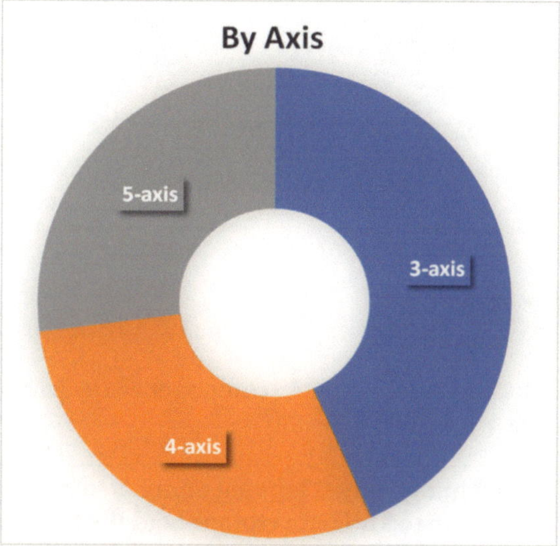

Figure 3.
Micromachining market share by axis.

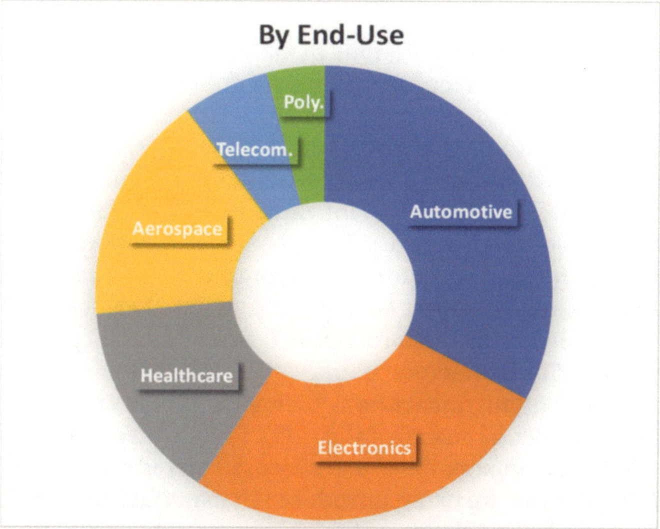

Figure 4.
Micromachining market share by end-use.

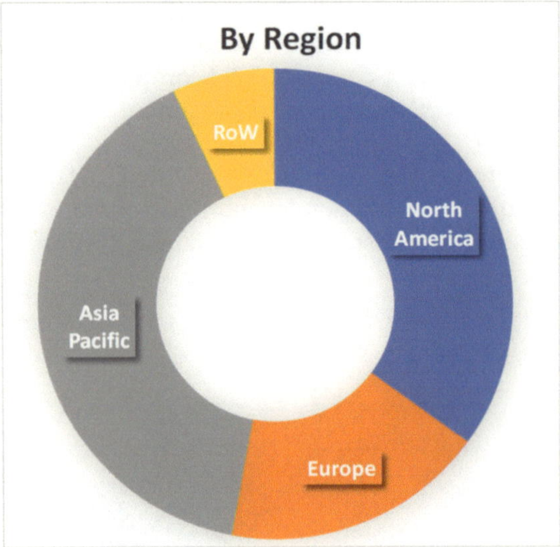

Figure 5.
Micromachining market share by region.

assemblies, sensors and optical MEMS components are driving the growth of military and aerospace industry.

- By region, micromachining market is being dominated by the Asia Pacific Region and North America (**Figure 5**). These regions hold the highest market share primarily because of the micromachined components (sensors, actuators, electrical assemblies, navigation controls, etc.) used in the automotive sector.

6. Conclusion

Although it is recognized as a milestone in manufacturing technology, providing excellent flexibility and high precision, micromachining technology still needs to be improved upon and developed. However, in comparison to other modern techniques, micromachining offers advantages like low production costs, small batch sizes, and the ability to manufacture precise 3D surfaces in a variety of materials. The ability of micromachining technology to reliably fabricate much smaller features at very tight tolerances is one of its main advantages. Even though lithography-based manufacturing can produce finer component features, micromachining still has many advantages over it in terms of produced geometry complexity, material choices and accuracy. Furthermore, the technology shows promise in terms of its ability to close the gap between the macro and nano/microdomains.

Another major asset of micromachining technology is the way micromachining intersects with environmental sustainability [25]. Precise machining reduces the material waste. Novel equipment designs are focused on reduced energy consumption. Eco-friendly practices are being introduced in manufacturing processes as well as a tendency towards the recyclability of micro components and waste materials.

As technology advances and its impact extends across various industries, micromachining will continuously work on reshaping the future of manufacturing by driving innovation and efficiency and redefining what is achievable in miniaturization and precision engineering.

Acknowledgements

The authors are grateful to the Ministry of Science, Technological Development and Innovations of the Republic of Serbia (Contract No. 451-03-66/2024-2103/200017) for the financial support.

Conflict of interest

The authors declare no conflict of interest.

Author details

Ivanka Stanimirović* and Zdravko Stanimirović
"Vinča" Institute of Nuclear Sciences, National Institute of the Republic of Serbia, University of Belgrade, Belgrade, Serbia

*Address all correspondence to: ivanka.stanimirovic@vin.bg.ac.rs

References

[1] Balázs BZ, Geier N, Takács M, et al. A review on micro-milling: Recent advances and future trends. International Journal of Advanced Manufacturing Technology. 2021;**112**:655-684. DOI: 10.1007/s00170-020-06445-w

[2] O'Toole L, Kang CW, Fang FZ. Precision micro-milling process: State of the art. Advanced Manufacturing. 2021;**9**:173-205. DOI: 10.1007/s40436-020-00323-0

[3] Airao J, Kishore H, Nirala CK. Measurement and analysis of tool wear and surface characteristics in micro turning of SLM Ti6Al4V and wrought Ti6Al4V. Measurement. 2023;**206**:112281. DOI: 10.1016/j.measurement.2022.112281

[4] Shrivastava A, Gangopadhyay S. Evaluation of adequacy of lubricants in MQL micro-drilling by a developed analytical model and experiments. Journal of Manufacturing Processes. 2023;**101**:1592-1604. DOI: 10.1016/j.jmapro.2023.07.017

[5] Batham H, Patel D, Banerjee T. A review on fabrication, grinding performance and failure of micro-grinding tools. Materials Today: Proceedings. 2022;**66**(9):3870-3877. DOI: 10.1016/j.matpr.2022.06.304

[6] Wang J, Fu J, Wang J, Du F, Liew PJ, Shimada K. Processing capabilities of micro ultrasonic machining for hard and brittle materials: SPH analysis and experimental verification. Precision Engineering. 2020;**63**:159-169. DOI: 10.1016/j.precisioneng.2020.02.008

[7] Ivanov A, Lahiri A, Baldzhiev V, Trych-Wildner A. Suggested research trends in the area of micro-EDM—Study of some parameters affecting micro-EDM. Micromachines. 2021;**12**(10):1184. DOI: 10.3390/mi12101184

[8] Rashid MAN, Saleh T, Noor WI, Ali MSM. Effect of laser parameters on sequential laser beam micromachining and micro electro-discharge machining. The International Journal of Advanced Manufacturing Technology. 2021;**114**:709-723. DOI: 10.1007/s00170-021-06908-8

[9] Li X, Ming P, Ao S, Wang W. Review of additive electrochemical micro-manufacturing technology. International Journal of Machine Tools and Manufacture. 2022;**173**:103848. DOI: 10.1016/j.ijmachtools.2021.103848

[10] Krimpenis AA, Noeas GD. Application of hybrid manufacturing processes in microfabrication. Journal of Manufacturing Processes. 2022;**80**:328-346. DOI: 10.1016/j.jmapro.2022.06.009

[11] Sahu AK, Malhotra J, Jha S. Laser-based hybrid micromachining processes: A review. Optics & Laser Technology. 2022;**146**:107554. DOI: 10.1016/j.optlastec.2021.107554

[12] Zh G, Li H, Zhang T, Liu M. Modeling and simulation of micro-hole fabrication on brittle material using abrasive air jet machining. Journal of Manufacturing Processes. 2021;**72**:361-374. DOI: 10.1016/j.jmapro.2021.10.036

[13] NSM A, Mahmoodian R, Hamdi Abd Shukor M. Effect of focused ion beam process parameter on tin-nickel-copper micropillars microfabrication. Materials and Manufacturing Processes. 2020;**35**(2):163-171. DOI: 10.1080/10426914.2020.1711923

[14] Gentili E, Tabaglio L, Aggogeri F. Review on micromachining techniques. In: Proceedings of the Seventh International Conference on Advanced Manufacturing Systems and Technology (AMST'05). Vienna: Springer; 2005. pp. 387-396. DOI: 10.1007/3-211-38053-1_37

[15] Gomes MC, da Silva MB, Duarte MAV. Experimental study of micromilling operation of stainless steel. The International Journal of Advanced Manufacturing Technology. 2020;**111**:3123-3139. DOI: 10.1007/s00170-020-06232-7

[16] Kieren-Ehses S, Böhme L, Morales-Rivas L, Lösch J, et al. The influence of the crystallographic orientation when micro machining commercially pure titanium: A size effect. Precision Engineering. 2021;**72**:158-171. DOI: 10.1016/j.precisioneng.2021.04.007

[17] Rawal S, Sidpara AM, Paul J. A review on micro machining of polymer composites. Journal of Manufacturing Processes. 2022;**77**:87-113. DOI: 10.1016/j.jmapro.2020.09.077

[18] Shanu A, Sharma P, Dixit P. Micromachining of alumina ceramic for microsystems applications: A systematic review, challenges and future opportunities. Materials and Manufacturing Processes. 2023;**39**(7):1-33. DOI: 10.1080/10426914.2023.2290244

[19] Geier N, Patra K, Anand RS, Ashworth S, et al. A critical review on mechanical micro-drilling of glass and carbon fibre reinforced polymer (GFRP and CFRP) composites. Composites Part B: Engineering. 2023;**110589**:1-21. DOI: 10.1016/j.compositesb.2023.110589

[20] Ranjan J, Patra K, Szalay T, Mia M, et al. Artificial intelligence-based hole quality prediction in micro-drilling using multiple sensors. Sensors. 2020;**20**(3):885. DOI: 10.3390/s20030885

[21] Zhao J, Jiang Z, Dobrzański LA, Lee CS, Yu F. Recent development in micromanufacturing of metallic materials. Materials. 2020;**13**(18):4046. DOI: 10.3390/ma13184046

[22] Micromachining Market—Global Forecast to 2025. Available from: https://www.marketsandmarkets.com/Market-Reports/mcromachining-market-167528458.html [Accessed: March 7, 2024]

[23] Micromachining Market—2020-2027 Forecast. Available from: https://www.grandviewresearch.com/industry-analysis/micromachining-market [Accessed: March 7, 2024]

[24] Global Micromachining Market Size, Trends & Analysis—Forecasts to 2026. Available from: https://www.globalmarketestimates.com/market-report/global-micromachining-market-2991 [Accessed: March 7, 2024]

[25] Schneider F, Das J, Kirsch B, et al. Sustainability in ultra precision and micro machining: A review. International Journal of Precision Engineering and Manufacturing-Green Technology. 2019;**6**:601-610. DOI: 10.1007/s40684-019-00035-2

Chapter 2

Micromachining and Its Applications for Electronics

Rakesh Kumar Bhardwaj and Anup Dutt

Abstract

Micromachining has emerged as a foundational technology in modern electronics, playing a crucial role in the creation of miniaturized and high-performance devices. Its significance is particularly pronounced in the domains of high-frequency RF devices and Terahertz communication, where the demand for precision, miniaturization, and performance enhancement is of utmost importance. The requirement of linear tolerances are of the order of 1–5 μm, surface toughness of 40–50 nm Ra and flatness of 0.3 μm.

Keywords: micromachining, tool based micromachining, electromechanical, terahertz (THz), waveguides

1. Introduction

The development of micro and nanotechnology has taken place due to the rising demand for micro and miniature parts and systems. The process used to fabricate MEMS and other microelectronics products can be called MEMS micro-manufacturing or lithography-based micro and nanomanufacturing. The technologies like photolithography, chemical etching, plating, and LIGA are called micro-manufacturing technologies. Micro manufacturing is generally employed to realize parts or feature sizes ranging from tens or hundreds of micrometers. Although micro manufacturing may not be enough to produce the smallest feature size as would be possible using MEMS and NEMS. This technology is a bridging gap for macro and nanodomain of manufacturing. It has various advantages over lithography-based micro-manufacturing in terms of material choice, relative accuracy, and complexity of part geometry. Micromachining is a broad term that encompasses a variety of fabrication processes for the production of miniature parts and structures. Non-lithography-based micromachining processes, such as micro electrical discharge machining (micro EDM), micro-milling, laser cutting/patterning/drilling, micro-extrusion, micro embossing, micro stamping, and micro injection molding, are widely used in the electronics industry. These processes employ diverse working principles and exhibit unique characteristics in terms of achievable accuracy, surface finish, and production rate. However, they all share the ability to produce three-dimensional microparts from a variety of materials [1]. This chapter will focus on micro mechanical cutting processes, in which the geometry of the cutting tool is defined. The relative accuracy and feature size, which can be achieved in micro-manufacturing, is described by Chang et al. [2] identified he has established that micro and nano manufacturing technologies is one of the most significant emerging manufacturing processes to address the future challenges in high-value engineering. Micromachining is a fabrication process that can be used to create three-dimensional (3D) microparts from a variety of materials, including metals, alloys, polymers, composites, and ceramics. Micromachined parts can have dimensions as small as 10^{-3} to 10^{-5} meters (sub-micron accuracy). This is significantly higher than the accuracy achievable with MEMS-based processes, which typically have accuracies of 10^{-1} to 10^{-3} meters [1].

A micro-cutting (micro-milling) has attracted growing attention from researchers and industry in the last 2–3 decades because mechanical cutting is a well-established area. Knowledge of macro cutting has been adapted to study the micro-cutting phenomenon. Two research approaches are being investigated; one is the minimization of the conventional cutting process, tooling, and equipment, with an emphasis on their scaling down effect. Micro cutting is kinematically similar to conventional cutting but fundamentally different from it in many aspects. Micro cutting refers to the mechanical micromachining (direct removal of metal) using geometrically defined cutter edge(s) on a conventional precision machine or micro machine. The selection of production method in microelectronics depends upon the number of replications and the degree of accuracy desired. Most of the production methods used nowadays are lithography based which are a combination of photolithography, etching and electrodeposition, and require costly setups and large volumes. The geometries obtained by these lithography-based processes are limited to 2½D microstructures with high aspect ratios and accuracies. Mechanical and thermal methods provide a viable solution for higher material removal rate during electronics manufacturing process. Such processes allow manufacturing of complex 3D shapes with acceptable aspect ratios, Bissacco et al. [3]. CNCmicro milling offers to process a wide range of materials at low set up costs. A large knowledge database and infrastructure developed for macro milling can be utilized for stepping into micro zone. Mechanical micromachining, particularly micromilling, offers several advantages over other manufacturing processes for electronics manufacturing due to its versatility, accuracy, and surface quality. Micromilling can be used to machine a wide variety of materials, generate complex 3D geometries with minimal setups, and produce parts with high accuracy and surface quality. However, micromilling also faces a number of challenges, including the size effect, rapid tool wear, inherent burr formation, low tool stiffness, and limitations on the minimum feature size and surface roughness compared to lithographic techniques [1]. The machining of micro molds require very high level of geometrical tolerances and surface quality this makes the application of micro end milling very challenging Aramcharoen et al., and Vázquez et al. [4, 5]. The generated topography of the micro milled surface is mainly influenced by the tool, machining parameters and cutting strategy. Micromachining enables miniaturization and integration of high-frequency RF devices and Terahertz communication, where size reduction is critical for efficient signal propagation Sankar et al. [6]. It also achieves precision fabrication of High-Frequency devices such as antennas, filters, and resonators which ensures optimal electromagnetic performance. The tight tolerances and fine features achievable with micromilling contribute to minimizing signal losses and maintaining the required device characteristics Kunieda et al. [7]. It also enables to tailor the shape and dimensions of resonators, waveguides, and other elements enables precise tuning of their electromagnetic properties, leading to enhanced device performance Mekonnen [8].

Micro cutting is kinematically similar to conventional cutting, but fundamentally different in many aspects [2]. Micro cutting refers to mechanical micromachining (direct removal of metal) using geometrically defined cutting edge (s) carried out on a conventional precision machines or micro machines. Aramcharoen et al. [4], Jain [9] has achieved and described stringent tolerances using tool based micro cutting. Wu et al. [10], Paulo et al. [11], Aramcharoen et al. [4] have described the size effect in micro milling the uncut chip thickness (UCT) is of the scale of cutting-edge radius or the tool nose radius. Based on the size effect Jun et al. [12] explained the elastic-plastic stage, which occurs due to plowing, which most of the workpiece material accumulates and expands ahead of the cutting edge. Relatively large burr formation occurs as a result of inhomogeneous distribution of material flows causing high surface roughness that decreases the material removal rate. Liu et al. [13] and Weule et al. [14] have given the material removal mechanism. In micro end-milling, the thickness initially increases from zero to a maximum value approximately equal to feed per tooth. After that it reduces from feed per tooth to zero. Whereas in the case of side milling, the thickness reduces from maximum to zero. This stage may occur repeatedly for a single point cutting edge, or in most common scenarios, simultaneously for multiple cutting edges depending upon the employed tool geometries, and the cutting parameters. Malekian et al. [15] predicted the minimum uncut chip thickness MUCT for Aluminum alloy (Al6061). Chae et al. [16], Sun and Cheng [17] and Chen et al. [18] discussed the importance of tool geometry and its effect on micro milling. Klocke et al. [19] studied the microstructure of cutting tool and suggested alloying effect on strength. Cheng et al. [20] studied the effect of cutting tool and emphasized the tool nose radius, rake and relief angle. Chen et al., Boswell et al., Yang et al. [21–23] described the importance of minimum quantity lubrication (MQL) for different materials based on experimental studies. In micromilling, the cutting forces are significantly influenced by factors that are typically negligible in conventional milling, such as tool wear, tool run-out, and chatteras described by Gietzelt and Eichhorn [24]. Tool run- out is more in micro milling due the low rigidity of the tool and higher cutting speeds, which leads to deteriorated surface [17]. Therefore to maintain tighter tolerances, reduce cutting force, maintain constant chip load, and avoid premature tool failure in micro scale milling, the cutting paths and machining strategies should be optimized [25, 26].

According to Griffin et al. [27] in the sub-millimeter or THz frequency range, the skin depth of the oscillating electric field in a metal is typically very small. This means that the electric field is only able to penetrate a very short distance into the metal. If there are defects or non-uniformities on the surface of the metal, these can act as additional resistance to the electrons. This is because the electrons must flow around the defects or non-uniformities, which takes longer and uses more energy. As a result, surface roughness is one of the most important factors affecting the performance of metal waveguides at sub-millimeter or THz frequencies. Surface roughness can cause signal losses and distortions. This is especially important in small waveguides, where the skin depth is even smaller and the effects of surface roughness are more pronounced. As per Narayanan et al. [28], the split-block technique is a conventional machining process that can be used to fabricate receiver components for the submillimeter and low terahertz frequencies. It involves machining the circuit structures on two (or more) metal blocks and then mating the blocks together to form the complete component. The split-block technique offers a number of advantages over other fabrication methods. First, it is relatively straightforward to implement. Second, it allows for the easy placement of circuit components such as RF chokes, diodes, and coupling structures on the split blocks prior to assembly. Zhang et al. [29], suggested strategies for reducing vibrations during micro milling. Quintana and Ciurana [30] presented chatter minimization strategies for suppression of chatter to get better surface finish. Researchers [31–35] discussed burr formation in details along with their location of occurrence. Strategies for minimization of burr formation were also discussed. Machine tool construction for micro machining and error sources were discussed by Lamikiz et al. [36] and Hashmi et al. [37] especially the Abbe's error and correction methods along with isolation from external vibration beds because of its very low thermal expansion coefficient of 6.5 μm/mK, 2.5 times less than the one of mineral casting. Additionally, granite with a density of 2.8 kg/dm^3 is beneficial. According to Uriarte et al. [38] we are using small tools; the spindle must rotate at high revolutions (<30,000 rev/min) to achieve the adequate cutting speed for most materials. Apart from the speed, the spindle must be stiff (>25 N/μm) and must present a small run out (<1 μm) to ensure high precision in the cutting process.

2. Micromachining techniques in electronics applications

2.1 Overview of micromachining techniques

Mechanical micromachining techniques, such as micromilling, micro–Electrical Discharge Machining (EDM), and laser micromachining, are indispensable tools in the creation of miniaturized electronic components. These techniques offer precise material removal and intricate feature generation, making them essential for fabricating compact and high-performance devices. Micro milling involves the use of rotating cutting tools to remove material from a work piece, creating intricate geometries with high precision. This technique is instrumental in the creation of miniaturized electronic components due to its accuracy, repeatability, and versatility in handling various materials. Micro milling enables the production of micro-scale features such as grooves, channels, and complex 3D structures. These structures are critical in microelectronics, MEMS devices, and micro-optics. Micromilling's role in miniaturized electronic components extends to applications like microfluidics, where precisely machined channels facilitate controlled fluid flow for lab-on-a-chip systems Câmara et al. [39]. Micro Electrical Discharge Machining (EDM) employs controlled sparks between an electrode and the workpiece to erode material and create intricate shapes. This non-contact process is pivotal in generating micro-scale features that are challenging to achieve with conventional machining methods. Micro EDM is used to fabricate micro-scale molds, tooling, and dies. These components play a crucial role in micro-injection molding for the mass production of microfluidic devices, micro-optics, and precision electronic connectors. The technique's ability to work with electrically conductive materials allows the production of miniaturized parts with high accuracy Kumar et al. [40]. Laser micromachining employs high-energy laser pulses to ablate or melt material, allowing the creation of intricate patterns with minimal heat-affected zones. This non-contact technique is versatile and suitable for various materials, including ceramics, metals, and polymers. Laser micromachining role in miniaturized electronics spans micro-via drilling in printed circuit boards (PCBs), creation of micro-optical components, and patterning for flexible electronics. Its precision and non-contact nature ensure minimal mechanical stress on delicate structures, making it an ideal choice for manufacturing micro-scale electronic components Gattass Rafael et al. [41].

DOI: http://dx.doi.org/10.5772/intechopen.113892

2.2 Micro milling, principle and process

2.2.1 Micro milling

Micro milling is a versatile micro-cutting process for producing micro-level components in small and medium quantities. CNC controllers allow for precise control of the cutting tool, while CAD/CAM software is used to generate the tool paths necessary to machine complex microstructures. The size, geometry, and composition of micro tools are all important factors in micro milling. The size of the tool determines the minimum feature size that can be machined. The geometry of the tool affects the cutting forces, surface finish, and tool wear. The composition of the tool affects the tool wear resistance and toughness. Commercially micro milling tools ranging from 25 to 1000 μm are available. A typical micro milling machine has got ultra-high-speed low run-out spindles for achieving acceptable machining rates. These spindles being temperature-controlled result in lower thermal errors and better surface finish. These machines use polymer concrete in place of cast iron for achieving high dynamic stiffness as well as submicron resolutions glass scales for better control and minimum motion errors (**Figure 1**). The machining of THz

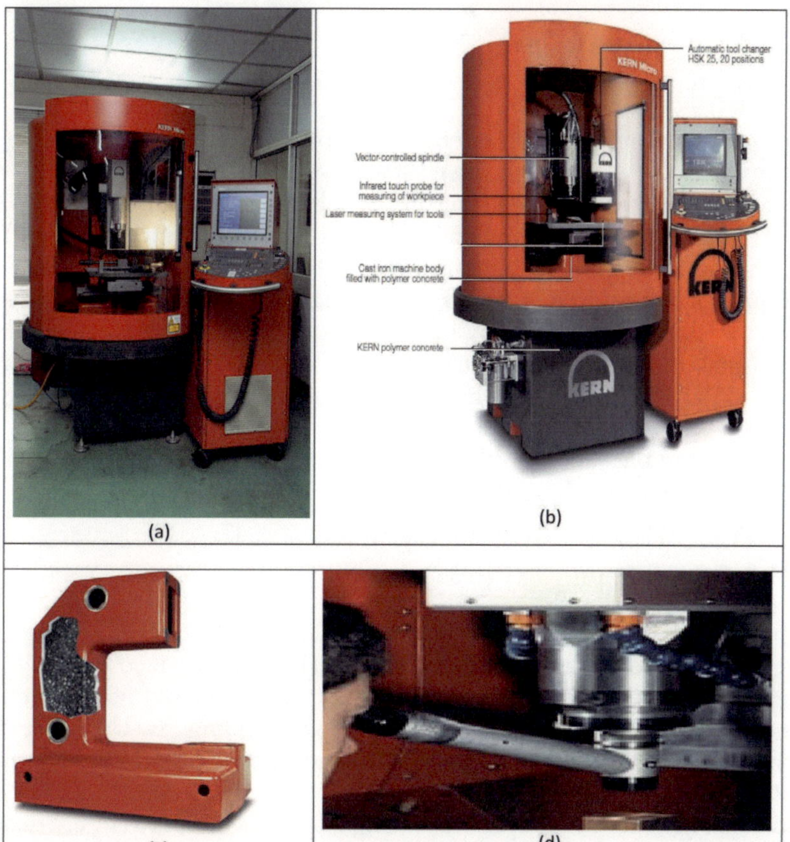

Figure 1.
(a) Kern Micro milling, (b) different systems of micro milling machine, (c) polymeric concrete construction, and (d) centring microscope.

Resource	Description
CAM system	Edge CAM and UG NX.
Machine tool	KERN Micro-2522, CNC Micromilling and drilling Machine, iTNC 530 Heidenhain Controller
Cutting tools	Ultrafine grade (<0.5 μm) WC-Co Solid Carbide Endmill Cutters
Cutting tool adapter	Balanced HSK 25 tool adapter and collet with runout less than 1 μm. (KERN-Schaublin make)
Cutting tool length and diameter calibration	Blum Micro Compact NT laser pre-setting system, repeatability ±1 μm
Workpiece holding	High Precision grinding vice
Configurations	Three linear axes
Base structure	Polymer concrete
Controller	Heidenhain iTNC 530
Speed range	Up to 50,000 rpm
Resolution	<0.1 μm
Work volume	250 × 220 × 200 mm

Table 1.
Features of a typical Micro milling machine.

waveguide geometries requires a very high level of geometrical tolerances and surface quality which can only be achieved through micro-milling [4, 5]. The generated topography of the micro-milled surface inside the waveguide is mainly influenced by the tool, machining parameters, and strategy of cutting. Hence micro-milling represents a suitable technique to manufacture THz waveguide components. **Table 1** shows a typical micro milling machine, and its features. This machine was extensively used to make waveguides at W band of frequency where the waveguide size is 2.54 × 1.27 mm within linear tolerances of 10 μm, surface roughness of the order of 150nmRa and flatness of 0.3 μm. when this frequent raises to 1THz the waveguide size becomes 0.245 × 0.127 mm within linear tolerances of 1–2 μm, surface roughness of 40–50 nmRa and flatness also of the order of half the wavelength at such a higher frequency Bhardwaj et al. [1, 42–46].

2.2.2 Cutting mechanisms

The material removal principles and process details of the micro-milling cutter are different from macro milling, and it can be characterized by small uncut chip thickness (UCT), specific cutting energy, tool run out noticeable size effect, and elastic recovery after machining. In micro-milling, the UCT is of the scale of the cutting-edge radius or the tool nose radius, also called the size effect (**Figure 2**) [11].

The unusual increase in the specific cutting energy as the depth of cut decreases below a critical value during micro-cutting operations using a geometrically defined tool is referred to as the size effect. It is among the principal issues in micro-milling which dominate cutting physics [12–17]. Three aspects of size-effect are as follows:

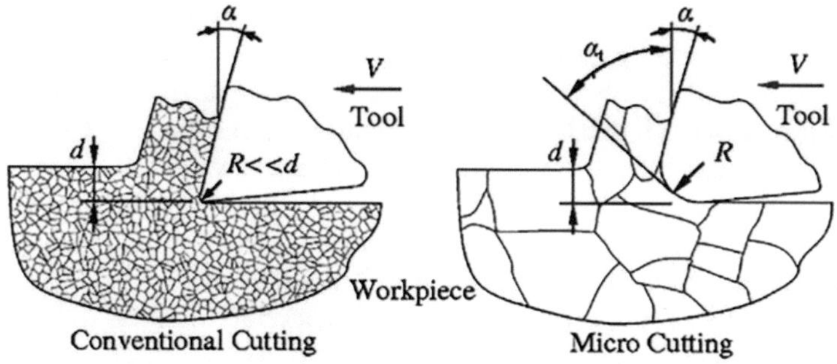

Figure 2.
Illustration showing the size effect in macro and micro-cutting [11].

i. If the uncut chip thickness is of the order of grain size, the work material may not exhibit homogeneous property.

ii. If the UCT is comparable to or smaller than the cutting-edge radius (tool nose radius). In that case, the rake angle becomes highly negative, resulting in the simultaneous occurrence of the plowing and shearing of the workpiece material.

iii. When the UCT in micro-cutting is smaller than a certain value, chip formation may not occur. This defines the minimum undeformed chip thickness (MUCT) hm.

As shown in **Figure 3a–c**, whether an elastic deformation or recovery of the machined surface occurs in the chip formation stage during micro milling depends on the chip thickness and the material of the workpiece. If the chip thickness is small enough, the workpiece material will only undergo elastic deformation and will fully recover to its original shape after the cutting tool passes. This is typically the case in the elastic stage of micro milling, where the chip thickness is in the order of a nanometer. If the chip thickness is larger than the critical chip thickness, the workpiece material will undergo plastic deformation and will not fully recover to its original shape after the cutting tool passes. However, some recovery may still occur, especially if the workpiece material is soft and ductile. This is typically the case in the elastic-plastic stage of

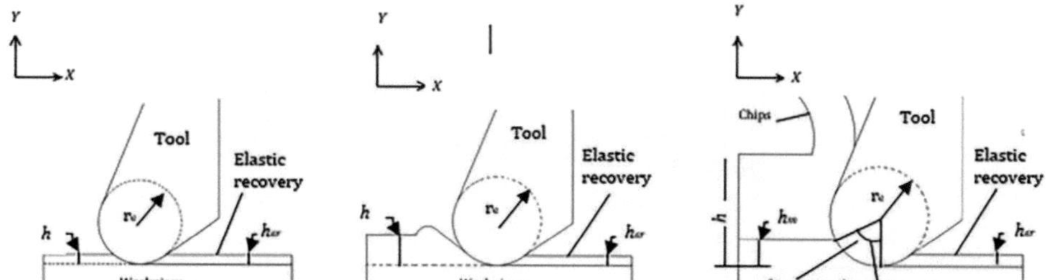

Figure 3.
Material removal in orthogonal micro-milling processes: (a) elastic deformation, (b) mixed elastic-plastic deformation region, and (c) complete chip formation region in the orthogonal micro-cutting process.

micro milling, where the chip thickness is in the order of a few micrometers to millimeters. Overall, it is important to note that there is no clear-cut boundary between elastic deformation and recovery of the machined surface during micro milling. The two phenomena are often intertwined and depend on a number of factors. However, a general rule of thumb is that the chip thickness and the material of the workpiece are the most important factors affecting the recovery of the machined surface [10, 13].

In the case of micro milling of THz waveguides, the chip thickness is typically very small, and the workpiece material is often a soft metal such as aluminum or copper. This suggests that elastic recovery of the machined surface is likely to be significant. However, more research is needed to quantify the exact amount of elastic recovery and to determine its impact on the performance of THz waveguides.

2.2.3 Material removal mechanism in micro-milling

In micro milling, the uncut chip thickness (UCT) initially increases from zero to a maximum value approximately equal to feed per tooth, and then decreases from feed per tooth to zero as shown in **Figure 4a**. In side milling, the UCT reduces from maximum to zero as shown in **Figure 4b**. The minimum uncut chip thickness (MUCT) is the smallest chip thickness that can be removed without causing plastic deformation of the workpiece material. The MUCT is based on the geometrical relationship at the stagnant point, which is the point at which the material deforms or flows with minimum deformation or kinematic energy [14]. The MUCT varies from 1/4 to 1/3 of the cutting-edge radius, depending on micro-milling conditions, micro mills (tools), and workpiece material classes [10]. Malekian et al. [15] predicted the MUCT for Aluminum alloy (Al6061) as $hm = re(1 - \cos\beta)$, where β is the friction angle between the workpiece and rake face. Son et al. [17] predicted the MUCT for Al, Brass, and OFHC as $hm = re(1 - \cos(\pi/4 - \beta/2))$.

The MUCT is an important parameter in micro milling, as it affects the material removal behavior, fluctuations in the cutting force, and stability of cutting forces. By understanding and controlling the MUCT, it is possible to improve the quality of the machined surface. The MUCT is influenced by a number of factors, including the geometry of the cutting tool, the material of the workpiece, and the micro-milling conditions. It is important to note that the MUCT is not a constant value, but rather varies depending on the specific micro-milling operation [18].

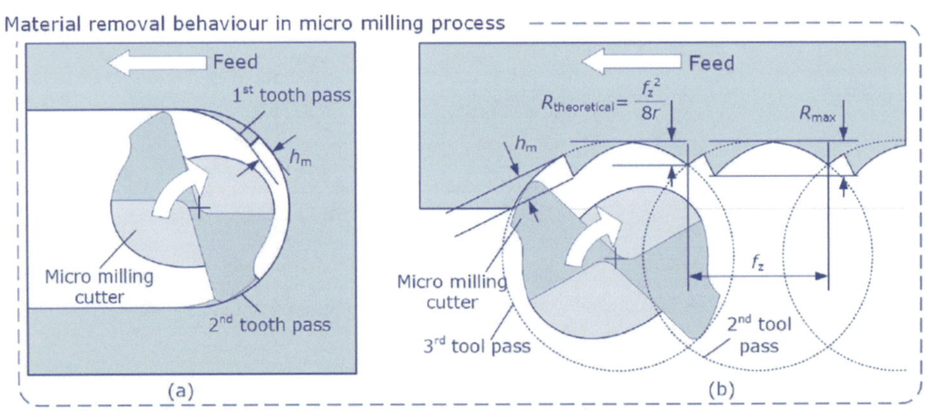

Figure 4.
(a) Tool pass in full end milling immersion, (b) side micro-milling process [18].

2.3 Cutting tools and their geometry effect on micro-cutting

Cemented carbide tools are an excellent choice for micro milling because of their high hardness, acceptable wear resistance, and strength, as well as their affordability. The performance of solid carbide tools is chiefly dependent upon the composition (i.e., cobalt content) and grain size. A lower grain size enables lower values of cutting-edge radius on the tool, which is important for achieving high surface roughness and accuracy in micro milling. The most popular choice for micro milling tools is fine grain size (<0.6 μm) cobalt bonded tungsten carbide [18–20]. Fine grain size tungsten carbide is an excellent choice for micro milling tools due to its high hardness, good wear resistance, high strength, and affordability. However, it is important to be aware of its brittleness and sensitivity to heat when using it.

The generic geometry of the micro cutting tool is shown in **Figure 5**, and raw material properties are given in **Table 2**.

The material properties of cemented carbides can be adjusted according to the requirement of micro milling cutter applications by altering cobalt content, average grain size, and concentration of composite carbides, as shown in **Figure 6a** and **b** [21].

2.4 Tool geometry and design considerations

Micro milling tools are small and delicate, and therefore they are more susceptible to breakage than larger tools. High stiffness is essential to prevent the tool from breaking, especially when machining hard materials. The optimal tool geometry for micro milling depends on the tool diameter. For tools with a diameter smaller than 200 μm, an asymmetrical tool geometry is preferred. This is because asymmetrical tools are more tolerant of fabrication errors. For tools with a diameter larger than 200 μm, symmetrical tool geometry is preferred. This is because symmetrical tools are more efficient. The relief angles, tool peripherals, and tool bottom surfaces should be

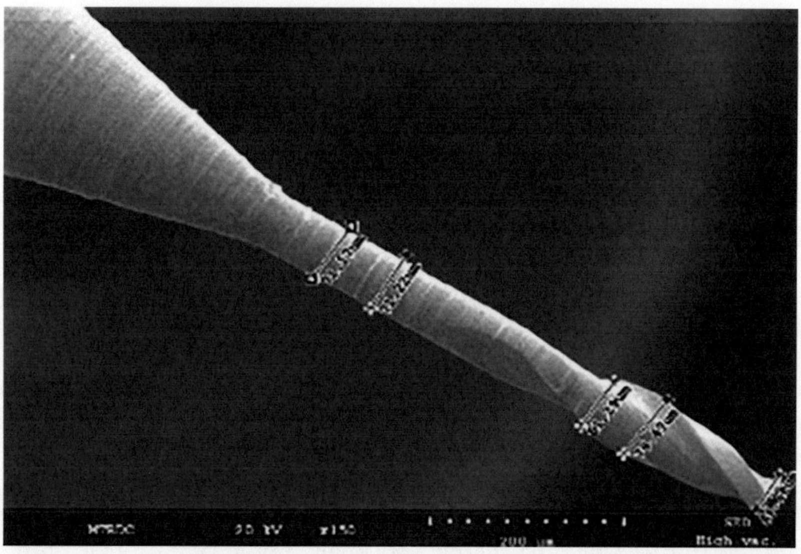

Figure 5.
SEM image of cutter having diameter of 70 μm [20].

Materials	Cemented carbide
Specification	HW-K10
Composition	WC-6Co
Density (g/cm^3)	14.90
Hardness (HV)	1580
Fracture toughness ($MPa.m^{1/2}$)	9.60
Flexural strength (GPa)	2
Compression strength (GPa)	5.40
Young's modulus (GPa)	630
Poisson constant	0.22
Heat conductivity ($W.m^{-1}.K^{-1}$)	80
Coefficient of thermal expansion, 293–1073 K, ($10^{-6}.K^{-1}$)	5.50
Specific heat capacity 293 K, ($J.g^{-1}.K^{-1}$)	0.95

Table 2.
Raw material types for micro milling cutters and their properties [20].

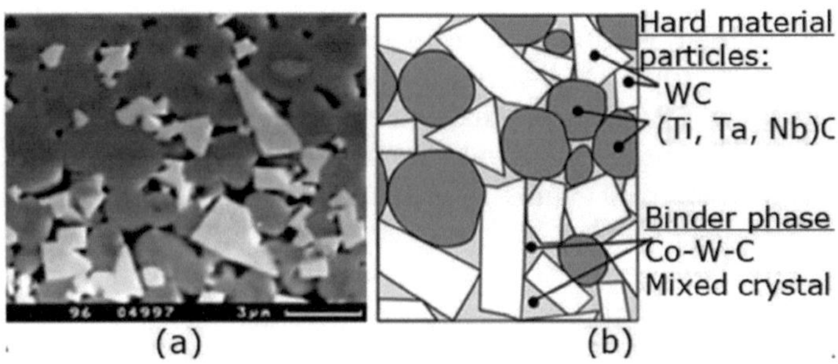

Figure 6.
(a) Scanning electron microscopy (SEM) images, (b) schematic of chemical components for micro-milling cutter materials [21].

carefully designed to avoid unnecessary contact with the workpiece. This is important to reduce friction and heat generation, which can lead to tool wear and poor surface finish. The critical small geometry features of the tool, such as the size of the smallest grinding wheels and thinnest electrode wire, must be considered during design. This is because these features can limit the minimum feature size that can be machined. Chip disposal spaces must be considered, especially when machining under severe cutting conditions such as no cutting fluid and high-pressure air. This is to ensure that the chips are removed from the cutting zone efficiently. Small cutting-edge radii are essential for achieving a high-quality machined surface. However, it is important to balance the need for a small cutting-edge radius with the need for wear resistance. The rake angle is an important parameter that affects the cutting forces, chip formation, and surface finish. A small positive rake angle is preferred for machining ductile materials. This is because a small positive rake angle produces lower cutting forces and a better surface finish. A highly negative rake angle is preferred for machining hard and brittle materials. This is because a highly negative rake angle produces higher cutting forces but a better surface finish. In addition to the criteria listed above, there are a number of other factors that need to be considered when designing micro milling tools, such as the material of the tool, the coating of the tool, and the cooling system. By carefully considering all of these factors, it is possible to design micro milling tools that can produce high-quality machined surfaces with high precision and accuracy [18, 22].

2.5 Minimum quantity lubrication (MQL)

Cutting fluid plays an important role because of its capacity to reduce friction and its ability to dissipate the heat generated between the micro tools and the workpiece. As cutting zones are very tiny, one of the most challenging aspects of the micro-milling scale is effective cooling and/or lubricating fluid delivery. More than 90% of the mechanical energy in micro-milling processes is converted into thermal energy, resulting in a temperature increase in the cutting zone and cutting edges. When machining oxygen-free copper and aluminum alloys, the cutting zone temperature is frequently below 100°C, whereas over 200°C is frequently what is observed when machining hard and brittle materials [23]. The conventional flood cooling method is not effective in micro-milling due to the small size of the cutting zone and the significant surface tension force [24, 25]. In order to improve cutting fluid penetration into the small cutting zone, a minimum quantity lubrication (MQL) method is considered as the suitable alternative for precisely delivering the lubricant and/or coolant and meeting the requirements of being environmentally friendly. In the MQL system, lubricant is pulverized, allowing to better reach the cutting edge. The turbulence created by tool rotation is a major challenge in micro-milling. It can lead to poor surface finish, tool wear, and chip formation. MQL systems help to overcome this challenge by delivering a small amount of lubricant to the cutting edge in the form of a mist. The mist is able to penetrate the turbulence and reach the cutting edge, where it can reduce friction, heat generation, and chip formation [26]. According to Li et al. [47], minimum quantity lubrication, also known as near dry machining (NDM), refers to the use of cutting fluids in tiny quantities, which is only about ten-thousandths of the amount of cutting fluid used in flood-cooled machining. MQL has been shown to be effective in improving tool life, surface finish, and burr formation in micro-milling.

2.6 Accuracy and thermal effects

Micromilling is significantly affected by factors that are often overlooked in conventional milling, such as tool wear, tool runout, and chatter [28]. Tool wear is high in micro-milling due to the lower chip loads and large effective rake angles. Tool runout is also higher in micro-milling due to the low rigidity of the tool and higher cutting speeds. Chatter is a common problem in micro-milling due to the low stiffness of the tool and the interrupted cutting process. Another challenge of micro-milling is that the stiffness of the tool decreases rapidly as the diameter of the tool decreases. This is because the tool can be considered as a cantilever, and the stiffness of a cantilever decreases with the fourth power of the diameter [29, 30]. Also, micro-milling is an interrupted cutting process where cutting forces vary with the rotation angle of the cutter. To overcome these challenges cutting paths and machining strategies should be selected and optimized in micro-scale milling to maintain tighter tolerances, reduce the cutting force, maintain constant chip load, and avoid premature tool failure [31, 32]. Gracia et al. [48] have discussed the sources of geometrical and surface errors affecting the machining accuracy in Micro milling (**Figure 7**). The key characteristics

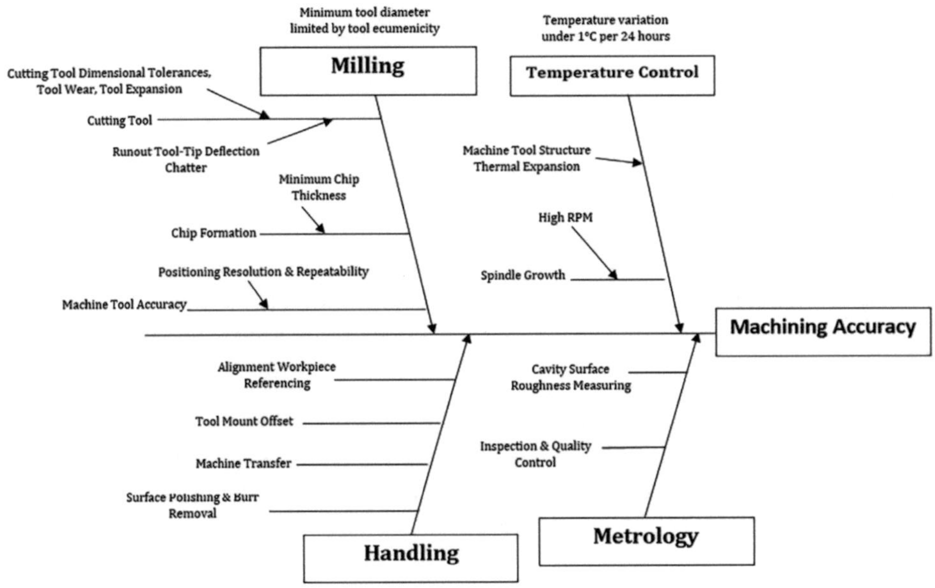

Figure 7.
Geometrical errors sources [31].

of micro-milling are shown in **Figure 8** in three broad categories i.e. Tooling, Work material and machine tool [11].

Micromachining processes for THz waveguide are particularly sensitive to the surface roughness and the precision of achieved machining geometrical accuracies. Apart from the inaccuracies of the setup and positional errors, another problem in CNC precision is that of achieved control versus desired control, where actuator backlash, wearing of gears can all add to the problems of deviations. Tool deflections occur in micro-milling, and vibrations occur based on the large tool length to small diameter ratios; such problems only add to increased geometrical inaccuracies. Finally, another problem to be considered is in terms of temperature gradients that can impact material machining accuracy.

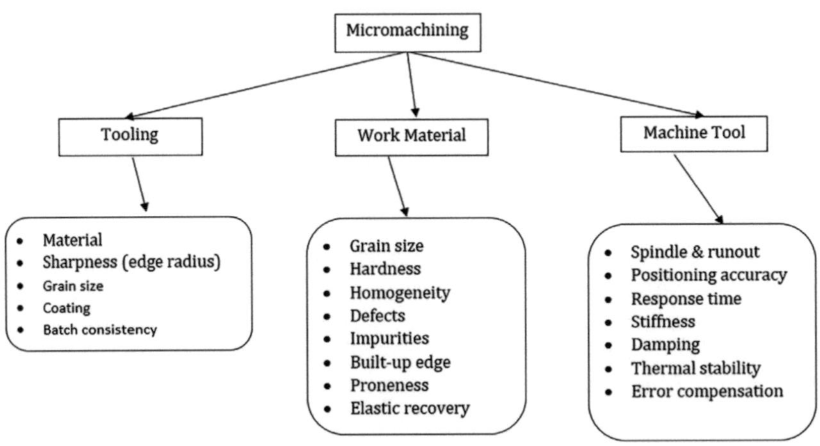

Figure 8.
Key characteristics in micro milling [11].

2.6.1 Higher spindle speeds (RPM) in micro-milling

High spindle speeds also have the effect of reducing the chip size by requiring a smaller feed rate per tooth, f_T (since $f_T \propto 1/N$). Smaller chip loads result in improved surface finish. Good surface finishes are crucial for cutting down on waveguide losses [33, 34].

2.6.2 Temperature control and thermal effects

The most important issue in precision machining is maintaining a constant temperature of the positional stages and the tool holders. Due to the stringent tolerances required for fabricating high-frequency waveguide components, temperature changes of a few degrees result in significant errors. To overcome the thermal effects, liquid cooled spindles and slideways are part of the precision machine design. As well as before commencing actual machining, the machine is usually run for 10–20 min to reach temperature equilibrium (warmup).

2.7 Chatter in machining

Chatter has been and still is a very important topic in micro-milling. It is a highly complex phenomenon due to the diversity of elements that can compose the dynamic system and its behavior. Negative effects of chatter are poor surface quality, unacceptable accuracies, excess noise, disproportionate tool wear, low material removal rate (MRR), material wastage, cost of rework or repair, wastage of energy, and tool damage. Chatter generates through the self-excitation mechanism. It can be primary chatter or secondary chatter.

Primary chatter is by friction between tool and work, it produces a thermomechanical effect on chip formation, and it is due to mode coupling. At the same time, *Secondary chatter* regenerates waviness on the workpiece surface. This is the main cause of chatter. In most publications, chatter is referred to as regenerative chatter. However, it is possible to distinguish between frictional chatter, thermomechanical chatter, mode coupling chatter, and regenerative chatter on the self-excitation mechanisms that cause the vibration [36].

2.7.1 Strategies for ensuring stable machining process

The control strategies are based on in-process control and out-of-process control. The first group is composed of all those methods that ensure a stable machining process by selecting cutting parameter combinations in the stable zone of the side lobe diagram (SDL) and making most of it. It is possible to distinguish between out of the process and in-process methods. The Methods aim to predict the location of the stability boundary of the cutting process to select stable cutting parameter combinations. The SLD identification is made out of the process before machining. On the other hand, In-process includes those methods that detect chatter during the metal cutting process, allowing the parameters to be corrected, so the cut migrates to the stable zone. It is possible to distinguish between passive and active mechanisms as follows:

2.7.2 Passive modes of chatter suppression

Strategies based on modifying certain machine tool elements to passively change the behavior of system composed of machine tool, for example, cutting tool and tool holder.

2.7.3 Active modes of chatter suppression

Based on certain elements capable of modulating the quantity of work provided, absorbing or supplying the energy with the aim to actively raise or at least change the stability frontier.

2.8 Burr formation

One of the main micro-milling-induced geometrical defects is the burr occurrence on the machined edges of the material. Burrs can be defined as the undesirable projections on the surface beyond the edge of the workpiece which are formed due to the bending of chips at the end of the cut. It has been reported and accepted that burr formation at the microscale is also affected with size effects analogous to the surface roughness.

When the ratio of cut to the cutting-edge radius is small, plowing dominates the cutting process rather than shearing, resulting in high biaxial compressive stresses that push material toward the free surface and generate large-top burrs. Also, the path the tool follows when entering or exiting the work surface also significantly affects burr formation [38, 39]. The cutting parameters, workpiece material properties, tool geometry, coatings, and coolant lubricants also affect the burr formation significantly micro-milling [41]. **Figure 9** shows different types of burr and their location.

Higher feed per tooth and smaller cutting width values have a positive effect on the burr formation mechanisms [42]. **Figure 10** gives the Ishikawa diagram is showing various possible factors which may affect the surface quality and burr formation of the micro-milled surface.

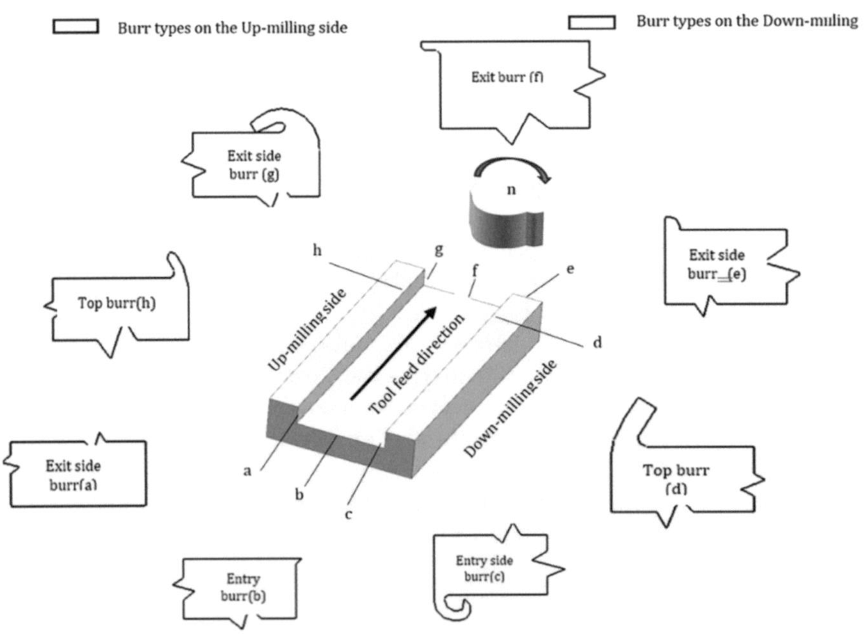

Figure 9.
Different burr types and their locations [38].

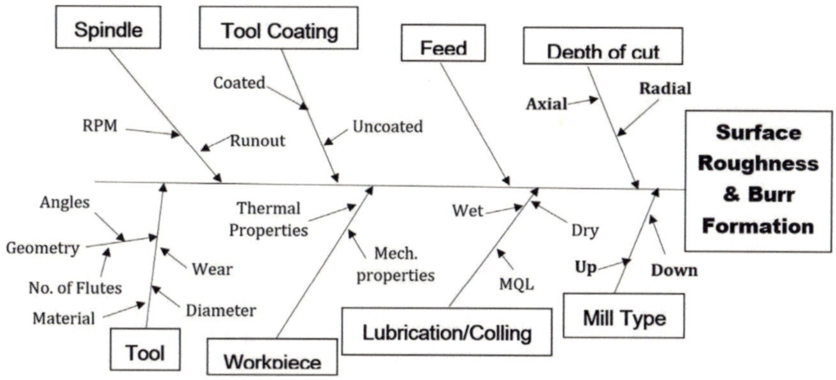

Figure 10.
Fish bone diagram of factors affecting surface quality.

2.9 Machine tools for micro-scale processing

Micro-scale processing is used to produce small and precise features in a wide range of industries, including automotive, aerospace, astronomy, medical, optics, and metrology. To meet the demands of these industries, it is necessary to develop machine tools that can provide the required accuracy and precision.

Figure 11 shows that a thermally and dynamically stable machine tool structure is important for micromachining because it helps to ensure that the workpiece is machined accurately and precisely. Thermal distortion and vibration can cause errors in the machining process, which can lead to defective products. New structural materials, such as advanced ceramics, advanced composite materials, engineering plastics, fiber-reinforced plastics (FRPs), and fiber-reinforced metals, can be used to improve the thermal and mechanical stability of machine tools for micromachining. **Figure 12** shows conceptual design of a thermally and dynamically stable machine tool. Design to minimize thermal distortion and vibration can be done by using materials with a low coefficient of thermal expansion (CTE), designing the structure to be stiff and rigid, and isolating the error sources from the machining area. By using vibration dampers and designing the structure to be symmetric and balanced, isolation of the error sources can be achieved. High-precision components, designing the structure to be easy to assemble and adjust, and using materials with a low CTE will lead to minimization of the error sources and feedback sensors will contribute to controlling of the error sources [49].

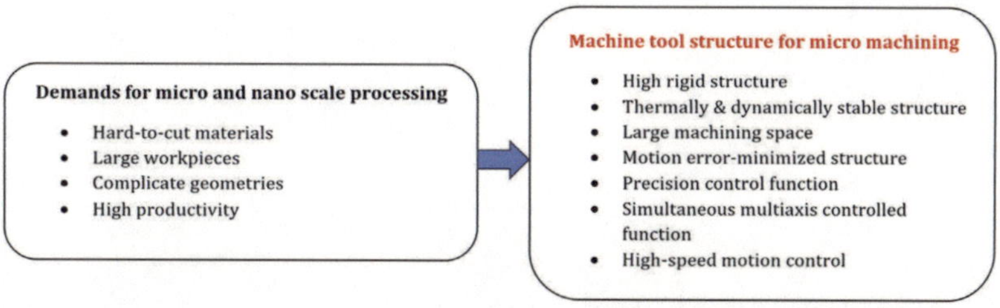

Figure 11.
Micro scale processing and machine tool structure.

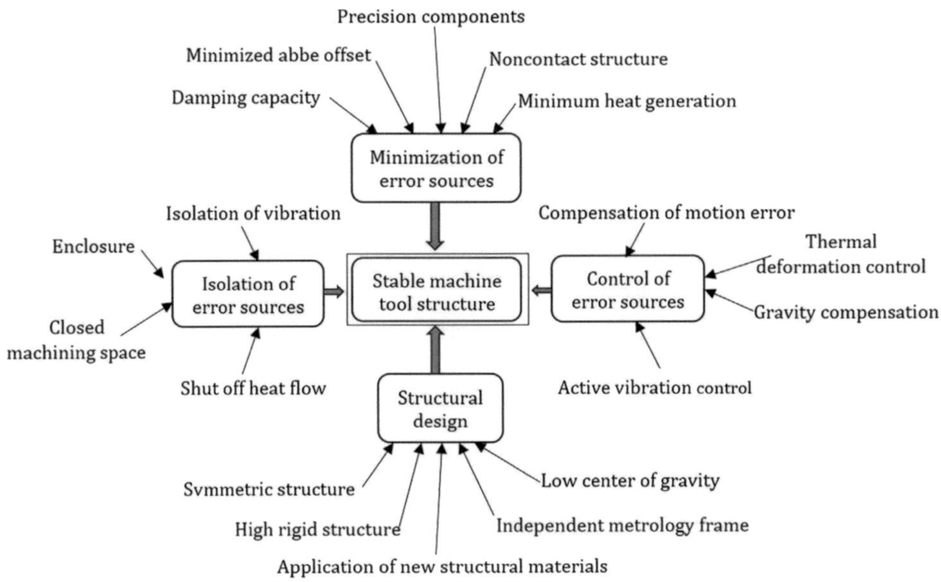

Figure 12.
Concept of stable machine tool structure for micro-and nanometer-scale processing [49].

2.9.1 Abbe's principle

The Abbe principle states that the length to be measured and the measuring scale must lie on the same axis. This is important because if the measurement is taken at an angle, it will introduce an error. Conventional machine tools often do not satisfy the Abbe principle because the motion axis and the machine element (its reference point or its center of gravity) do not lie on a single line. This can lead to errors in the machining process. **Figure 13** shows two different machine tool designs. **Figure 13b** shows a C-frame with a single ball screw column. In this design, the tool axis line and measuring line are separated. This means that the Abbe principle is not satisfied, and angular deflection will be translated into a displacement error at the tool. **Figure 13a** shows a double-column and twin ball screw

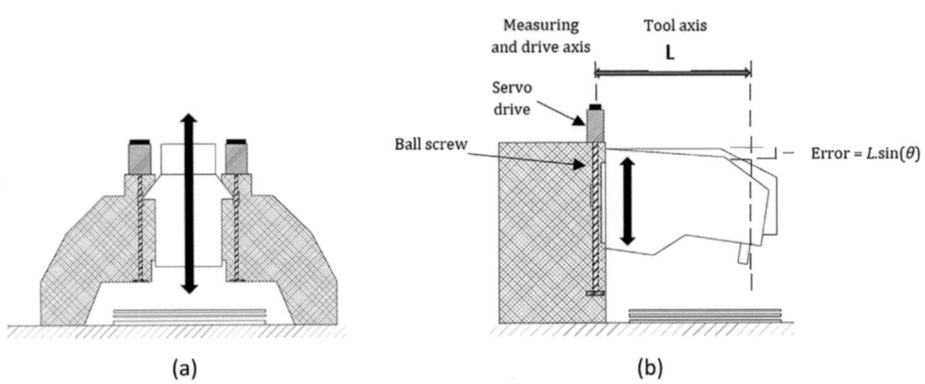

Figure 13.
Two different machine designs. (a) with Abbe's error reduction, (b) with Abbe's error.

design. In this design, the distance between the tool axis and each ball screw axis compensates for the angular deviation effect on the linear axis. The Abbe principle is an important consideration for the design of machine tools and other precision instruments. By following the Abbe principle, engineers can design machines that are more accurate and precise.

2.9.2 Machine bed material

Thermal stability is important for machine tool beds in micro-cutting because it helps to ensure that the machine bed maintains its shape and dimensions even when it is subjected to heat from the machining process. This can be achieved by using materials with a low thermal expansion coefficient, low specific heat capacity, and good damping properties. Granite and mineral casting are two materials that meet these requirements, and they are commonly used for machine tool beds in micro-cutting applications. Granite is used mostly in the field of ultra-precision machine beds because of its low thermal expansion coefficient of 6.5 μm/mK which is nearly 2.5 times less than the one of mineral casting. Additionally, granite with a density of 2.8 kg/dm^3 is beneficial in terms of Eigen frequency and dynamic properties (refer **Table 3**).

Mineral casting materials are a good alternative to granite for machine tool beds because they have similar material properties, but they are less expensive and easier to manufacture. The material properties of mineral casting are comparable to those of granite, but the heat expansion coefficient is about two to three times higher. However, the thermal distortion of mineral casting machine tool beds can be minimized by using appropriate design techniques and by carefully controlling the temperature of the machine tool environment [45].

2.10 Touch-trigger probing and laser measuring systems

Probing systems are an indispensable part of micromachining. Probing systems can help to improve the accuracy, efficiency, and quality of manufactured products. On-machine probing is used to measure and inspect workpieces while they are still on the CNC machine tool. This can help to reduce setup and inspection times and improve the overall efficiency of the machining process. In-process probing is used to monitor the machining process and to make adjustments as needed. This can help to improve the accuracy and quality of the finished product.

Characteristic	Unit	Grey cast iron	Granite	Mineral cast
Compression strength	N/mm^2	600–1000	70–300	140–170
Tensile strength	N/mm^2	150–400	30–35	25–40
Modulus of elasticity	kN/mm^2	80–120	35–90	30–40
Heat conductivity	W/mK	50	1.7–2.4	1.3–2.0
Coefficient of thermal expansion	m/mk	10	6.5–8.5	12–20
Density	g/cm^2	7.15	2.9–3.0	2.1–2.4
Damping capacity	—	0.003	0.015	0.02–0.03

Table 3.
Material properties for structural elements in precision machine tools.

2.10.1 Application of probing measuring

Standard applications in micromachining of touch-trigger, on-machine probing measuring systems, has following benefits:

- Identification of the type and size of parts, and their alignment in the fixture.
- Simplification of traditional fixtures by coupling a pallet fixturing system with machine/tool probes because the probe locates everything and compensates before machining.
- Tool presetting and blunt or broken tool detection.
- Checking tools, verification of setups and adjusting offsets.
- Routine machine calibration.

3D touch probes and laser controlled, in process tool management systems are two types of non-contact tool measurement systems that can be used to improve the accuracy and efficiency of CNC machining processes.3D touch probes (as shown in **Figure 14**) are mounted in the machine spindle and can be used for a variety of tasks, including workpiece alignment, datum setting, workpiece measurement, and digitizing 3D surfaces. Laser controlled (as shown in **Figure 15**), in process tool management systems use lasers to measure tool length, diameter, roundness, and run-out with a repeatability of better than 1 µm. These systems can also be used to detect tool wear and to compensate for thermal drift of the machine CNC axis and tool.

2.11 Precision vices

Some features of Precision vices are, use of high-grade FCO 60 body, ground top and bottom. The clamping jaws give high performance without wear and accuracy

Figure 14.
3D touch probe, RENISHAW OMP.

Figure 15.
Laser controlled, in process tool management system.

in parallelism and squareness of the order of 0.02/100 mm. the jaw plates should be interchangeable. Tenon and Tenon slots are provided for effortless trueing of vice on machine table, high clamping force of the order of 5000 Kg can be achieved effortlessly so that hammering is eliminated on vice handle.

2.12 Spindle/tool holders/collets

Spindles in micro-milling machines must be able to rotate at high speeds (up to 30,000 rpm) to achieve the adequate cutting speed for most materials. The spindle must also be stiff (> 25 N/μm) and have a small runout (< 1 μm) to ensure high precision in the cutting process. To achieve high speeds, spindles in micro-milling machines typically use ceramic ball bearings that are continuously refrigerated and lubricated. Micro-milling spindles are also typically low-power electro-spindles (200–500 W) or aerostatic spindles. Tools in micro-milling are typically small and delicate. The most used collets for clamping tools in micro-milling machines are precision "ER type" collets and super-precision ER type collets. Precision ER type collets can present big runout errors that depend on the clamped diameter, while super-precision ER type collets present runout errors smaller than 2 μm. Tool wear in micro-milling is quite high, so it is usual to use two or more mills per operation (one for rough machining and one for finishing). Tool change is a critical operation in micro-milling because the tool runout, tool height, and collet runout are modified. Therefore, tool change must be performed carefully, cleaning all the shank, collets, tools, and nuts and applying controlled torques [46].

2.12.1 HSK dual-contact tooling

The HSK tool holder is a popular alternative to the conventional steep-taper spindle connection for high-speed machining (HSM). HSK tool holders offer several advantages, including:

- Simultaneous fitment on both the taper and face at the front of the spindle: This provides increased rigidity of the joint and improves accuracy.

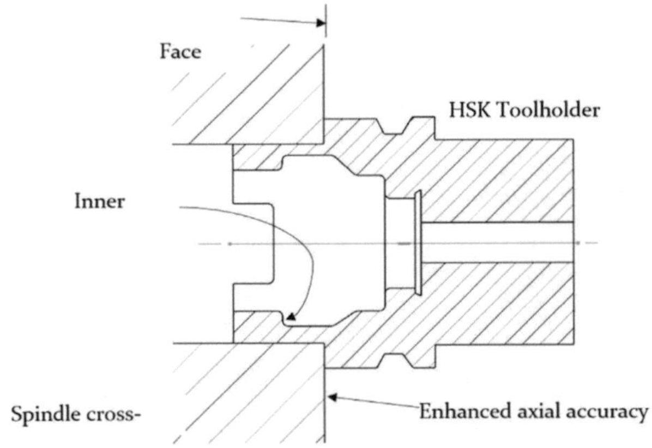

Figure 16.
HSK, dual-contact tooling.

- Reduced dimensions: HSK tool holders are smaller and lighter than conventional steep taper tool holders, which reduces vibration and improves performance at high speeds.

- Inner clamping: HSK tool holders clamp the tool shank from the inside, which provides better grip and reduces the risk of tool slippage (**Figure 16**).

3. Challenges and advances in micromilling for RF devices

3.1 Challenges in micromilling RF devices

Micromilling has revolutionized the fabrication of high-frequency RF (Radio Frequency) devices, enabling miniaturization and enhanced performance. However, this process comes with challenges that researchers and engineers have been addressing through innovative advancements.

3.1.1 Challenges

1. *Dimensional accuracy and surface finish*: Micromilling demands high dimensional accuracy and surface finish for precise RF device performance. Achieving sub-micron tolerances while maintaining a smooth surface in the micro-scale features is challenging due to tool wear, vibration, and thermal effects [18].

2. *Tool wear and material removal rates*: Maintaining tool sharpness and durability during micromilling is critical. The small tool sizes used for micromilling can experience rapid wear, affecting accuracy and material removal rates. As RF devices require intricate structures, tool wear management becomes crucial [50].

3. *Vibration*: Microscale structures are susceptible to vibration effects, impacting accuracy and surface quality. These challenges become more pronounced as device sizes decrease and operating frequencies increase.

DOI: http://dx.doi.org/10.5772/intechopen.113892

3.2 Advances and solutions

1. *Tooling advancements:* Advances in tool design, materials, and coatings have led to longer tool life and improved machining performance. Ultra-hard tool materials, such as diamond-coated micro end mills, enhance wear resistance and enable prolonged micromilling operations [18, 51, 52].

2. *Process monitoring and control*: Real-time monitoring and feedback systems have been developed to detect tool wear, surface quality deviations, and other process anomalies. These systems enable adjustments during micromilling to maintain accuracy and quality [53, 54].

3. *Simulation and modeling:* Advancements in simulation software have enabled accurate prediction of tool behavior, cutting forces, and surface quality during micromilling. These models aid in optimizing tool paths, reducing vibration, and improving dimensional accuracy [55, 56].

4. *Multi-axis micromilling:* Multi-axis micromilling machines enable the fabrication of complex 3D structures without requiring multiple setups. These systems enhance efficiency, reduce errors, and enable precise fabrication of intricate high-frequency RF device geometries [57, 58].

5. *Hybrid micromachining*: Combining micromilling with other micromachining techniques, such as micro EDM and laser micromachining, provides a comprehensive solution to overcome challenges like tool wear and material compatibility. This approach enhances the accuracy and versatility of the micromachining process [59, 60].

4. Conclusion

The challenges and advancements in micromilling of high-frequency RF& Terahertz devices illustrate the ever-changing nature of this field. While challenges like dimensional accuracy, tool wear, and material compatibility still persist, innovative tooling, monitoring systems, simulation software, multi-axis systems, and hybrid approaches are advancing at rapid ratesto and increasing micromilling capabilities. These advances will continually enhance the precision and quality of high-frequency RF & TeraHertz devices, driving progress in modern communication and sensing systems.

Author details

Rakesh Kumar Bhardwaj* and Anup Dutt
Defence Electronics Applications Laboratory, Dehradun, Uttrakhand, India

*Address all correspondence to: bhardwaj9274@gmail.com

References

[1] Bhardwaj RK, Dutta VP, Bhatnagar N. Mechanical engineering challenges in micromachining of terahertz waveguide components. In: El Ghzaoui M, Das S, Lenka TR, Biswas A, editors. Terahertz Wireless Communication Components and System Technologies. Singapore: Springer Nature; 2022. pp. 231-258. DOI: 10.1007/978-981-16-9182-9

[2] Kai Cheng DH. Micro-Cutting: Fundamentals and Applications. 1st ed. Chichester: John Wiley & Sons Inc; 2013

[3] Bissacco G, Hansen HN, De Chiffre L. Micromilling of hardened tool steel for mould making applications. Journal of Materials Processing Technology. 2005;**167**(2):201-207

[4] Aramcharoen A, Mativenga PT. Size effect and tool geometry in micromilling of tool steel. Precision Engineering. 2009;**33**(4):402-407

[5] Vázquez E, Amaro A, Ciurana J, Rodríguez CA. Process planning considerations for micromilling of mould cavities used in ultrasonic moulding technology. Precision Engineering. 2015;**39**:252-260

[6] Sankar R, Li X. Micromachining technologies for microsystem integration and applications. Micromachines. 2016;**7**(4):62

[7] Kunieda M, Suzuki N. High-precision micro-machining technology. CIRP Annals. 2015;**64**(2):773-796

[8] Mekonnen AG, Dhar S. Design and modelling of micro machined terahertz resonators for label-free biomolecular detection. Journal of Micromechanics and Micro Engineering. 2017;**27**(1):015007

[9] Jain VK. Introduction to Micromachining. India: CR Press; 2001

[10] Wu X, Li L, He N, Yao C, Zhao M. Influence of the cutting edge radius and the material grain size on the cutting force in micro cutting. Precision Engineering. 2016;**45**:359-364. DOI: 10.1016/j.precisioneng.2016.03.012

[11] Paulo Davim MJJJ. Nano and Micromachining. Hoboken, USA: Wiley Online Library; 2009

[12] Jun MBG, DeVor RE, Kapoor SG. Investigation of the dynamics of microend milling—Part II: Model validation and interpretation. Journal of Manufacturing Science and Engineering. 2006;**128**(4):901-912. DOI: 10.1115/1.2335854

[13] Liu X, DeVor RE, Kapoor SG. An analytical model for the prediction of minimum Chip thickness in micromachining. Journal of Manufacturing Science and Engineering. 2005;**128**(2):474-481. DOI: 10.1115/1.2162905

[14] Weule H, Hüntrup V, Tritschle H. Micro-cutting of steel to meet new requirements in miniaturization. CIRP Annals – Manufacturing Technology. 2001;**50**(1):61-64. DOI: 10.1016/S0007-8506(07)62071-X

[15] Malekian M, Mostofa MG, Park SS, Jun MBG. Micromachining of minimum uncut chip thickness in micro machining of aluminium. Journal of Materials Processing Technology. 2012;**212**(3):553-559

[16] Chae J, Park SS, Friheit T. Investigation of manufacturing

micro-cutting operations. International Journal of Machinery Tool and Manufacturer. 2006;**46**(3-4):313-332

[17] Sun X, Cheng K. Chapter 2 – Micro-/Nano-machining through mechanical cutting. In: Qin Y, editor. Micro-Manufacturing Engineering and Technology. Boston: William Andrew Publishing; 2010. pp. 24-38. DOI: 10.1016/B978-0-8155-1545-6.00002-8

[18] Chen N et al. Advances in micro milling: From tool fabrication to process outcomes. International Journal of Machine Tools and Manufacture. 2021;**160**:103670. DOI: 10.1016/j.ijmachtools.2020.103670

[19] Klocke F. Manufacturing Processes 1. Berlin, Heidelberg, Germany: Springer; 2011

[20] Cheng X, Wang Z, Nakamoto K, Yamazaki K. A study on the micro tooling for micro/nano milling. The International Journal of Advanced Manufacturing Technology. 2011;**53**(5):523-533. DOI: 10.1007/s00170-010-2856-3

[21] Chen G, Ren C, Zhang P, Cui K, Li Y. Measurement and finite element simulation of micro-cutting temperatures of tool tip and workpiece. International Journal of Machine Tools and Manufacture. 2013;**75**:16-26. DOI: 10.1016/j.ijmachtools.2013.08.005

[22] Boswell B, Islam MN, Davies IJ. A review of micro-mechanical cutting. The International Journal of Advanced Manufacturing Technology. 2018;**94**(1):789-806. DOI: 10.1007/s00170-017-0912-y

[23] Yang K, Liang Y, Zheng K, Bai Q, Chen W. Tool edge radius effect on cutting temperature in micro-end-milling process. The International Journal of Advanced Manufacturing Technology. 2011;**52**(9):905-912. DOI: 10.1007/s00170-010-2795-z

[24] Gietzelt T, Eichhorn L. Mechanical Micromachining by Drilling, Milling and Slotting. IntechOpen; 2012. DOI: 10.5772/34124

[25] Qin Y. Micromanufacturing Engineering and Technology. 2nd ed. Waltham, USA: Elsevier; 2015

[26] Kuram E, Ozcelik B. Effects of tool paths and machining parameters on the performance in micro-milling of Ti6Al4V titanium with high-speed spindle attachment. The International Journal of Advanced Manufacturing Technology. 2016;**84**(1):691-703. DOI: 10.1007/s00170-015-7741-7

[27] Griffin JM et al. Control of deviations and prediction of surface roughness from micro machining of THz waveguides using acoustic emission signals. Mechanical Systems and Signal Processing. 2017;**85**:1020-1034. DOI: 10.1016/j.ymssp.2016.09.016

[28] Narayanan G, Erickson NR, Grosslein RM. Low Cost Direct Machining of Terahertz 1 Introduction. Amherst, USA. 1999. pp. 519-529

[29] Zhang SJ, S. To, Zhang GQ, Zhu ZW. A review of machine-tool vibration and its influence upon surface generation in ultra-precision machining. International Journal of Machine Tools and Manufacture. 2015;**91**:34-42. DOI: 10.1016/j.ijmachtools.2015.01.005

[30] Quintana G, Ciurana J. Chatter in machining processes: A review. International Journal of Machine Tools and Manufacture. 2011;**51**(5):363-376. DOI: 10.1016/j.ijmachtools.2011.01.001

[31] Lekkala R, Bajpai V, Singh RK, Joshi SS. Characterization and modeling

of burr formation in micro-end milling. Precision Engineering. 2011;**35**(4):625-637. DOI: 10.1016/j.precisioneng.2011.04.007

[32] Aurich JC, Dornfeld D, Arrazola PJ, Franke V, Leitz L, Min S. Burrs-analysis, control and removal. CIRP Annals – Manufacturing Technology. 2009;**58**(2):519-542. DOI: 10.1016/j.cirp.2009.09.004

[33] Balázs BZ, Geier N, Takács M, Davim JP. A review on micro-milling: Recent advances and future trends. The International Journal of Advanced Manufacturing Technology. 2021;**112**(3):655-684. DOI: 10.1007/s00170-020-06445-w

[34] Kumar P, Kumar M, Bajpai V, Singh NK. Recent advances in characterization, modeling and control of burr formation in micro-milling. Manufacturing Letters. 2017;**13**:1-5. DOI: 10.1016/j.mfglet.2017.04.002

[35] Piquard R, D'Acunto A, Laheurte P, Dudzinski D. Micro-end milling of NiTi biomedical alloys, burr formation and phase transformation. Precision Engineering. 2014;**38**(2):356-364. DOI: 10.1016/j.precisioneng.2013.11.006

[36] Lamikiz A, López de Lacalle LN, Celaya A. Machine Tools for High Performance Machining. Springer; 2009. pp. 219-260

[37] Hashmi S. Comprehensive Materials Processing. Waltham, USA: Elsevier; 2014

[38] Uriarte L, Eguia J, Egaña F. Micromilling machines. In: López de Lacalle LN, Lamikiz A, editors. Machine Tools for High Performance Machining. Springer; 2009. pp. 369-397

[39] Câmara MA, Campos Rubio JC, Abrão AM, Davim JP. State of the art on micromilling of materials, a review. Journal of Materials Science & Technology. 2012;**28**(8):673-685

[40] Kumar D, Singh NK, Bajpai V. Recent trends, opportunities and other aspects of micro-EDM for advanced manufacturing: A comprehensive review. Journal of the Brazilian Society of Mechanical Sciences and Engineering. 2020;**42**:222. DOI: 10.1007/s40430-020-02296-4

[41] Gattass Rafael R, Mazur E. Femtosecond laser micromachining in transparent materials. Nature Photonics. 2008;**2**(4):219-225

[42] Bhardwaj RK, Sudhamani HS, Dutta VP, Bhatnagar N. Micromachining and characterisation of folded waveguide structure at 0.22THz. Journal of Infrared, Millimeter, and Terahertz Waves. 2021;**42**(3):229-238. DOI: 10.1007/s10762-021-00767-w

[43] Rakesh Bhardwaj VP, Bhatnagar N. Mechanical engineering challenges in making terahertz technology next frontier in wireless communication. Defense Technology Spectrum. 2015;**102**:98-102

[44] Sudhamani HS, Bhardwaj R, Reddy SUM, Balakrishnan J. Study of passband and stopband properties of sheet-beam folded waveguide structure. IEEE Transactions on Electron Devices. 2019;**6**(5):2401-2408

[45] Rakesh Bhardwaj VP, Bhatnagar N. Micromachining of Teraguides. National Journal of Mechanical Engineering, Space society of Mechanical Engineering. 2016;**26**:2015

[46] Bhardwaj RK, Dutta VP, Bhatnagar N. Material selection, micromachining, and measurements of THz waveguide components. In:

El Ghzaoui M, Das S, Lenka TR, Biswas A, editors. Terahertz Wireless Communication Components and System Technologies. Singapore: Springer Nature; 2022. pp. 259-292. DOI: 10.1007/978-981-16-9182-9

[47] Li KM, Chou SY. Experimental evaluation of minimum quantity lubrication in near micro-milling. Journal of Materials Processing Technology. 2010;**210**(15):2163-2170. DOI: 10.1016/j.jmatprotec.2010.07.031

[48] García G, Siller HR, Medrano-téllez A. Manufacturing guidelines for micro injection molds. In: International Conference on Micro Manufacturing (ICOMM), No. 8. 2013. pp. 658-664

[49] Yoshioka H, Shinno H. Structural design of a newly developed ultraprecision machine tool "ANGEL". International Journal of Automation Technology. 2011;**5**:38-44

[50] Zhu K, Yu X. The monitoring of micro milling tool wear conditions by wear area estimation. Mechanical Systems and Signal Processing. 2017;**93**:80-91

[51] Richter B, Morrow JD, Martin B, Heaney PJ, Segersten J, Pfefferkorn FE. Comparing the performance of diamond coated and uncoated micro end mills using robustness testing. Procedia CIRP. 2014;**14**:360-365

[52] Sousa VF, Silva FJ. Recent advances on coated milling tool technology—A comprehensive review. Coatings. 2020;**10**(3):235

[53] Hsieh WH, Lu MC, Chiou SJ. Application of back propagation neural network for spindle vibration-based tool wear monitoring in micro-milling. The International Journal of Advanced Manufacturing Technology. 2012;**61**:53-61

[54] Malekian M, Park SS, Jun MB. Tool wear monitoring of micro-milling operations. Journal of Materials Processing Technology. 2009;**209**(10):4903-4914

[55] Moges TM, Desai KA, Rao PVM. Modeling of cutting force, tool deflection, and surface error in micro-milling operation. The International Journal of Advanced Manufacturing Technology. 2018;**98**:2865-2881

[56] Mamedov A. Micro milling process modeling: A review. Manufacturing Review. 2021;**8**:3

[57] Huo D, Cheng K, Wardle F. Design of a five-axis ultra-precision micro-milling machine—UltraMill. Part 1: Holistic design approach, design considerations and specifications. The International Journal of Advanced Manufacturing Technology. 2010;**47**:867-877

[58] Bang YB, Lee KM, Oh S. 5-axis micro milling machine for machining micro parts. The International Journal of Advanced Manufacturing Technology. 2005;**25**:888-894

[59] Chavoshi SZ, Luo X. Hybrid micro-machining processes: A review. Precision Engineering. 2015;**41**:1-23

[60] Kunar S, Singh AK, Raviteja D, Kibria G, Chatterjee P, Perveen A, et al. Overview of Hybrid Micromachining and Microfabrication Techniques. Hybrid Micromachining and Microfabrication Technologies: Principles, Varieties and Applications. Beverly, USA. 2023. pp. 1-15

Section 2

Micromachining Techniques

Chapter 3

Advancement in Micromachining Techniques of MEMS Piezoresistive Pressure Sensors to Minimize Offset Drift Due to Humidity and Temperature

Bagur Rudrappa Deepu, Belthangady Pavithra,
Mangalore Manjunatha Nayak and Kunchinadka Narayana Bhat

Abstract

The present chapter focuses on the micromachining technology advancement to enhance the silicon micromachined piezoresistive pressure sensor's output characteristics. The differential pressure sensors with a glass bond or a passivation layer of Cr/Au metals at the device's rear side are fabricated using MEMS techniques. The sensors with modifications can minimize and almost eliminate the absorption/adsorption of moisture content on the sensor surface for a long time, indicating improved device performance. The effect of atmospheric temperature and humidity on the four piezo resistors and the sensor's drift is investigated in this chapter, highlighting the challenges involved in glass micromachining techniques, including wet etching, sandblasting, and electrochemical discharge machining (ECDM). Ultimately, the fabricated sensor's pressure calibration and offset drift values are studied due to atmospheric effects being reported.

Keywords: MEMS, differential piezoresistive pressure sensor, passivation layer, glass micromachining, sandblasting, electrochemical discharge machining, sensor's drift

1. Introduction

Pressure sensors based on Micro Electro Mechanical Systems (MEMS) technology are devices that combine micromechanical structures and electronic components integrated into silicon chips. Most pressure transducers have a diaphragm as the sensing element, which deflects under pressure. This mechanical deflection ultimately gives the electrical output according to the transduction mechanism *viz*, capacitive, piezoresistive, or piezoelectric methods [1]. The piezoresistive property, which was discovered back in 1920, gained significant application in silicon-based transducers in the 1950s [2, 3]. Piezoresistive-based transducers find vast applications over other

MEMS devices due to their linear output over a wide range of pressure and batch fabrication processes. Since the invention of MEMS pressure transducers, many studies have focused on addressing issues with drift, accuracy, sensitivity, nonlinearity, and hysteresis. As a result, we have a matured MEMS technology today. The applications of piezoresistive-based pressure sensors are receiving close attention due to their wide variety of applications in biomedical, defense, industrial control, automotive, and aerospace applications.

According to several publications [4], sensors can experience drift due to humidity, temperature (thermal), and residual stresses on the diaphragm caused by fabrication and packaging. Many efforts have been made to improve the pressure sensor's characteristic properties. Coating an n-type shield layer over p-type piezo resistors to counteract drift caused by electrical fields and temperature [5]. SOI (Silicon On Insulator) wafer-based pressure sensors came into existence to get a very low temperature coefficient of gauge factor [6, 7]. Piezoresistive pressure sensors fabricated from wideband-gap semiconductors such as GaN, diamond, and SiC-based pressure sensors were used for extreme temperature environments [8–10]. Polycrystalline silicon-based pressure sensors were developed to overcome thermal drift issues [11]. Via etch technology and surface micromachining technologies were explored in depth for improved pressure sensor characteristics [12, 13].

The atmospheric humidity and temperature adversely affect device performance, which leads to the misinterpretation of the real sensor's output/data. Even though the necessary electronic compensation can effectively (or somewhat) reduce this effect, an advancement in the sensor microfabrication process is investigated here.

Since the invention of transistors, glasses have been used as sealings which protect the device against external moisture and other contamination [14, 15]. Researchers have extensively used glass substrate which is bonded anodically to the MEMS absolute pressure sensors using an evacuated bonding system [16]. Whereas this step is overlooked due to the fabrication complications for low pressure MEMS differential pressure sensors. This chapter focuses on addressing those fabrication difficulties and highlights the importance of glass bonding in sensor's drift. Besides, an alternate method of Cr/Au passivation layer on the rear side of the sensor is introduced to counteract the drift issues of low pressure (0–140 mbar) differential MEMS pressure sensors.

1.1 Working principle of piezoresistive pressure sensors

Piezoresistive pressure sensors consisting of a silicon diaphragm with four resistors connected in a Wheatstone bridge configuration. When the pressure is applied, the diaphragm undergoes mechanical deformation as shown in **Figure 1a**. As a result, the piezo resistors (R_1, R_2, R_3 and R_4) on the diaphragm also change their resistance values accordingly. The proportional change in resistance is being converted into electrical output through a Wheatstone bridge configuration, as shown in **Figure 1b**. **Figures 2** and **3** represents the fabricated MEMS pressure sensor die.

2. COMSOL analysis and design considerations

Generally, piezoresistive pressure sensors are constituted by a silicon diaphragm on which four resistors are located at high-stress regions. The measuring method followed in such sensors is a Wheatstone bridge. Before the sensor fabrication process, finding the diaphragm dimension and the placement of the four piezo resistors on

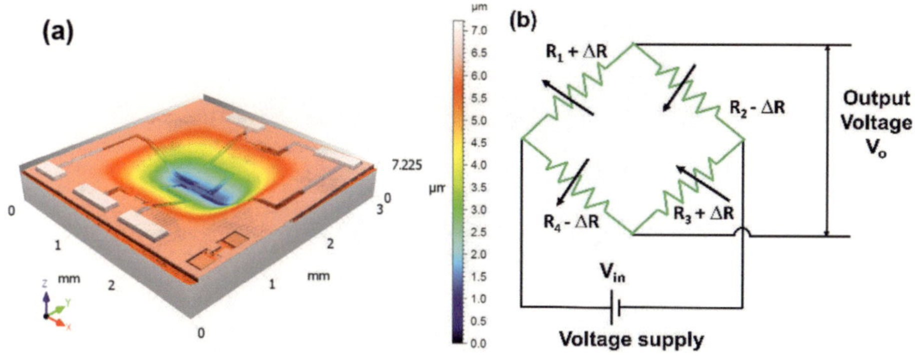

Figure 1.
(a) Optical profilometric image showing the mechanical deformation of the diaphragm upon applied pressure and (b) piezo resistors in Wheatstone bridge configuration.

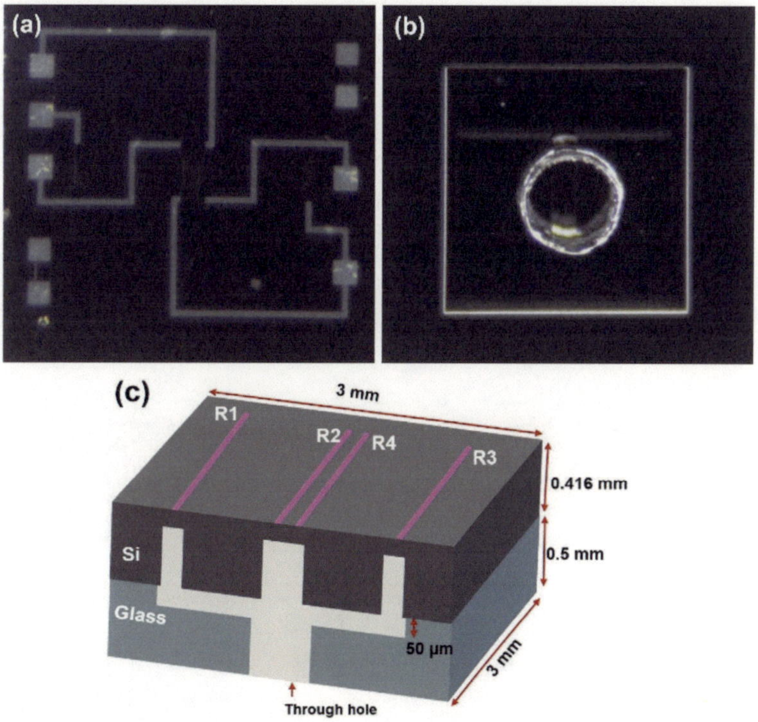

Figure 2.
MEMS differential pressure sensor die (a) front view, (b) rear view and (c) schematic view.

the diaphragm to get a high-sensitivity region is a prerequisite. COMSOL simulations analyze the piezo-resistors' position and the diaphragm's dimensions. The detailed analysis and design considerations are well reported in [1]. The fabrication of a differential pressure sensor in this study, wherein the diaphragm (2 × 2 mm) with a double BOSS structure (0.8 × 0.8 mm) called a sculptured diaphragm was considered to get a linear output at a small pressure range of 0–140 mbar. The diaphragm is anchored all around its edges, and the piezo-resistors are placed at the high-stress region as shown in **Figure 3**. The rigorous COMSOL analysis (**Figure 4b**) of the micromachined

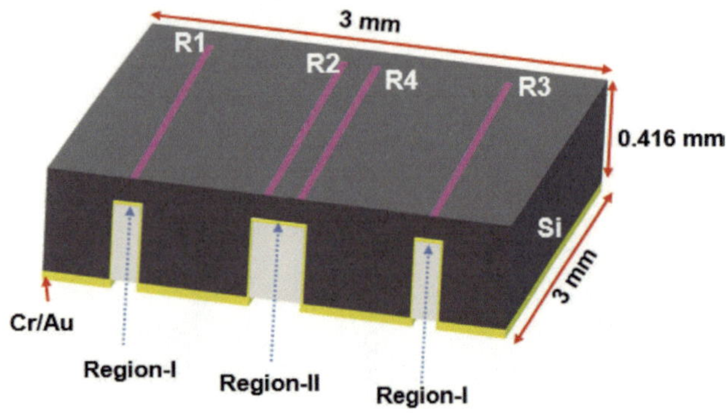

Figure 3.
Schematic view of double BOSS structured differential MEMS pressure sensor.

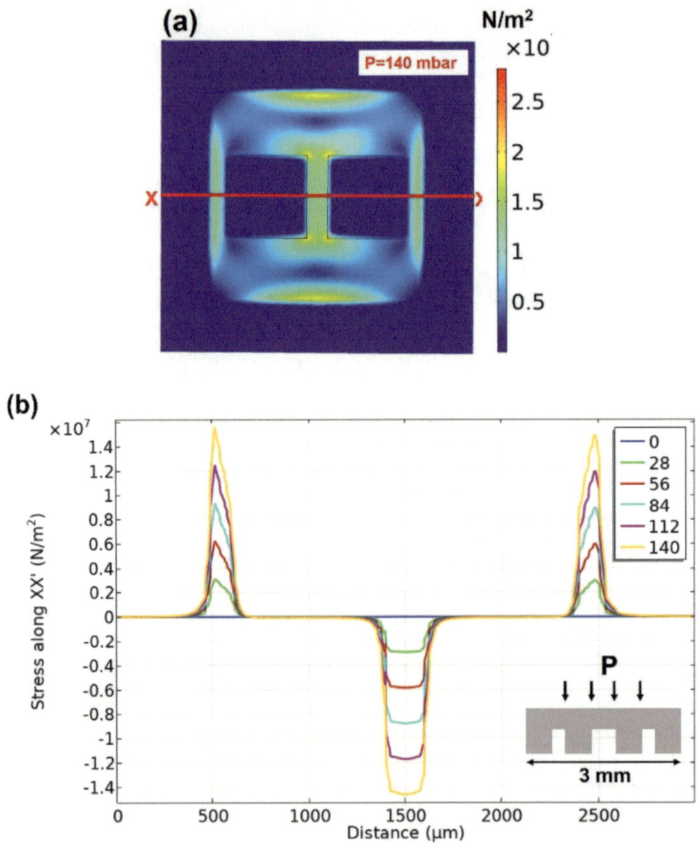

Figure 4.
COMSOL simulation indicating (a) stress distribution along XX' for 140 mbar applied pressure and (b) the stress distribution on the top surface of the diaphragm (inset) for varying magnitude of applied pressure P (in mbar) = 0, 28, 56, 84, 112, and 140.

double BOSS diaphragm estimates the position of the transverse stress distribution along XX' (marked in **Figure 4a**). The location of the maximum stress is usually localized at thin areas of the diaphragm.

It can be seen from **Figures 3** and **4a** and **b** that all four resistors placed at regions I and II which undergo transvers stress upon the application of input pressure. When the pressure is applied onto the diaphragm, the resistors R1 and R3 placed at region-I undergo identical transverse tensile stress and are connected to the opposite arms of the Wheatstone bridge. Conversely, the resistors R2 and R4 placed at region-II undergo transverse compressive stress and are connected to the other opposite arms of the bridge. From the simulation, the average stress values were found to be 0.016 GPa for transverse tensile stress and − 0.016 GPa for transverse compressive stress. The negative sign indicates the compressive stress.

3. Designing of pressure sensors by Clewin software

Piezoresistive MEMS pressure sensor fabrication involves seven photo lithographic patterning steps. As per the simulation, the photomask layout for the sensor fabrication was drawn using Clewin software, as shown in **Figure 5**, indicated by different colors for the respective layers. The fabrication involves five mask-layer processes for SOI wafer and two mask-layer processes for glass wafer.

4. Fabrication of MEMS pressure sensors

The pressure sensor dice were fabricated on 4-inch SOI substrates at Centre for Nano Science and Engineering (CeNSE), Indian Institute of Science (IISc), Bangalore, India, as shown in **Figure 6**. Four lithographic steps on frontside of the SOI wafer were performed defining the P+ region, piezo resistors, metal contacts

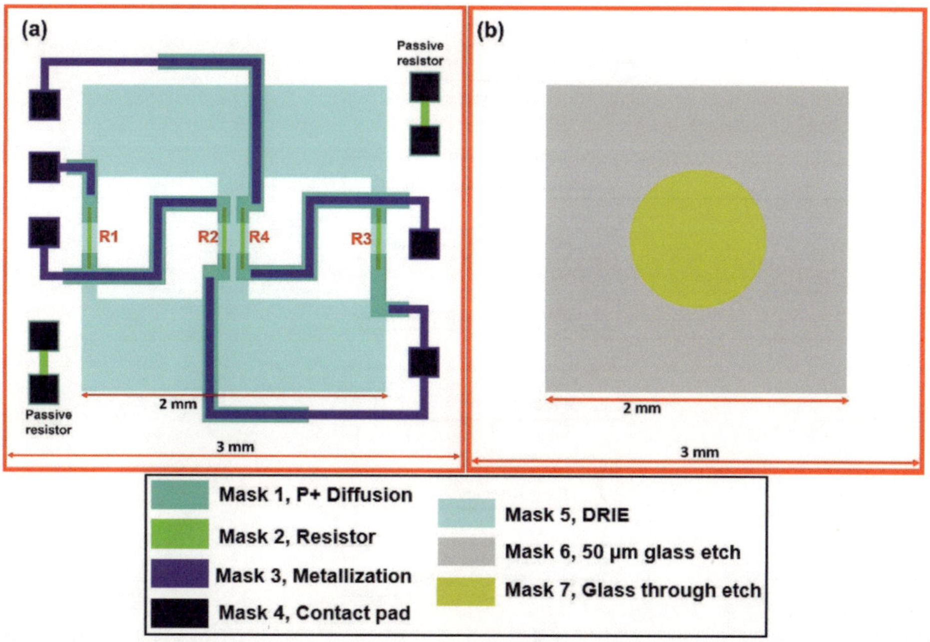

Figure 5.
Clewin layout for the fabrication process on (a) 4-inch SOI wafer and (b) 4-inch glass wafer.

for the open Wheatstone bridge configuration, and the passivation layer. The fifth lithographic step was carried out to release the diaphragm from the wafer backside by dry etching. The detailed process steps are described as follows.

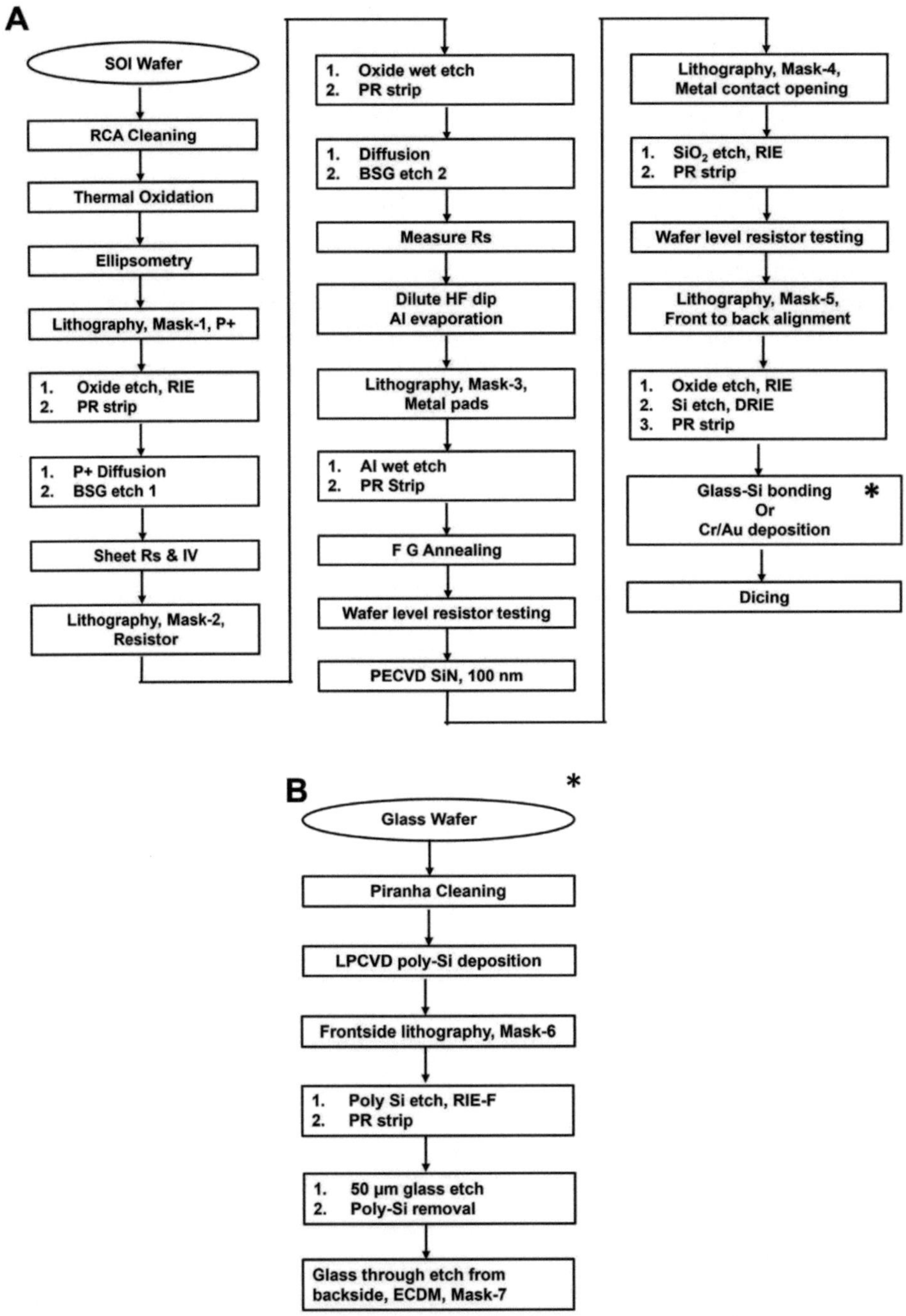

Figure 6.
(A) Steps for the fabrication of differential MEMS pressure sensors and (B) glass wafer fabrication process flow.

4.1 Fabrication process on SOI wafer

The fabrication process was commenced by thermally oxidizing the SOI (100) wafer with an n-type device layer of 15 μm, a buried oxide layer of 1 μm, and p-type handle layer of 400 μm thickness. Photomask-1 was used to define the P+ region by etching the oxide layer using buffered hydrofluoric acid (BHF) solution. Boron doping was carried out to get 5–6 Ω/square sheet resistance at 1100°C for 1 h. The four resistors were patterned using photomask-2, and BHF solution was used to etch the oxide layer. B-doping was performed to obtain sheet resistance of ~300 Ω/square at 1100°C for 10 min. Photomask-3 was used to define the metal contacts followed by the SiN passivation layer deposition by PECVD. The SiN film on the metal pads was etched off to define the area for wire bonding (photomask-4). Finally, front-to-back alignment lithography and DRIE were performed to release the diaphragm. **Figure 6A** shows the fabricated SOI wafer steps.

4.2 Glass bond and metal passivation on rear side of SOI wafer

From the fabrication flow chart (**Figure 6A**), two approaches were considered to study the pressure sensor's drift issues. Approach 1 is the glass bonding, and approach 2 is the Cr/Au deposition at the rear side of the SOI wafer as follows.

4.2.1 Glass etch and anodic bonding (approach 1)

Figure 6B describes the fabrication steps for glass etch and anodic bonding. Before anodic bonding, the glass (SD32 HOYA glass) was etched to a depth of 50 μm (photomask-6) on one side by the wet etch method. 32% hydrofluoric acid and 3.2% hydrochloric acid as an etchant, and 1 μm polycrystalline Si (poly-Si) deposited by LPCVD was used as an etch mask. This etch is necessary for the safe deflection of the BOSS structured diaphragm upon applied pressure range. Through hole is made on glass from the other side by ECDM technique (photomask-7). The different glass micromachining techniques are discussed in detail in Section 5. Processed SOI and glass wafers were bonded successfully using an anodic bonding method at 1100 V DC, 400°C for 10 min and 10^{-4} mbar vacuum. The fabricated wafer was diced (**Figure 7**) using the dicing tool (DISCO).

4.2.2 Metal passivation layer (approach 2)

Cr/Au thin film of 40/200 nm was deposited on fabricated SOI wafer rear side using thermal evaporation at a base pressure of 10^{-5} mbar as passivation layer.

5. Glass machining techniques

Glass and quartz are extensively used in MEMS, biochip fabrication, and microfluidic devices due to their excellent material properties such as mechanical robustness, transparency, hardness, biocompatibility, chemical and heat resistant nature. The hermetic sealing of glass is required for the absolute pressure sensors, which is usually bonded anodically in a high vacuum. Though the glass bonding in the case of differential pressure sensors is not necessary, it is an added advantage if included. Glass bond provides mechanical strength, enables high electrical insulation for the

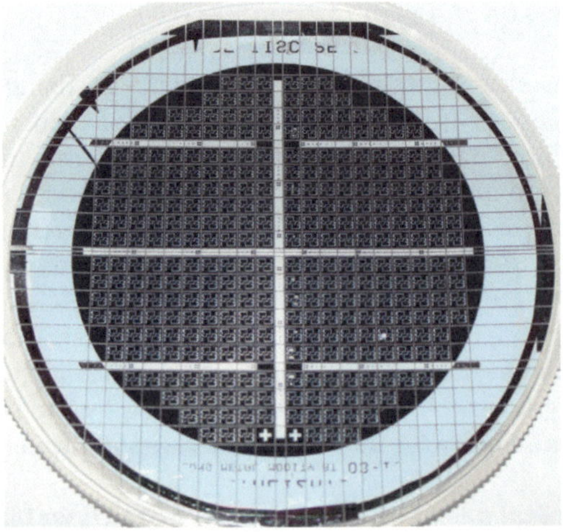

Figure 7.
Photograph of the fabricated 4-inch SOI wafer after dicing.

sensor from the Kovar header (on which the sensor is mounted using compatible epoxy material), protects the die from the external environment, and enhances the lifetime. Importantly, the temperature effects are minimized as the thermal expansion coefficient of glass matches with Kovar material. Unfortunately, every additional fabrication step is accompanied by some process complexities. Glass machining has many challenges that must be carefully addressed. Various conventional glass etch techniques include mechanical methods such as ultrasonic drilling, powder/sand blasting, and electrochemical micromachining *etc*. As mechanical machining has limitations below 100 μm, microfabrication techniques such as dry plasma etch, and wet chemical etch are preferred. A careful selection of glass micromachining techniques according to the applications is a crucial step.

This work considers three glass patterning techniques for differential MEMS pressure sensor fabrication. Before the anodic bonding, the glass machining of ~50 μm depth from one side and a through hole from the other side must be carried out using photomask 6 and 7 (**Figure 6b**). The glass was bonded anodically to silicon with an excellent hermeticity. However, the surface roughness and cleanliness of the wafer surface to be bonded are highly important factors for the fabrication yield. The prerequisite for the best glass etching is a compatible mask and the lowest possible effects on the etched surface. Higher surface roughness (> 20 nm) results in anodic bonding failures. The challenges faced in each of the machining techniques are discussed as follows.

5.1 Glass pattern by wet etching and sand blasting

The growing electronic industry demands cost-effective and efficient fabrication technology. The advantages and disadvantages of glass machining by wet etch, sandblasting and ECDM are compared in **Table 1**. Irrespective of isotropic etch nature, the wet etch received much attention compared to others due to its process simplicity, high mask selectivity, low surface roughness, and cost-effectiveness. As far as the through glass etch is concerned, the wet method is accompanied by

Parameter	Wet etching	Sandblasting	ECDM
Etch depth	350 μm	500 μm	500 μm
Lateral etch	Yes (~1:1)	No	No
Aspect ratio	Low	High	High
Process time	Less (2 h)	Less (30 min)	Moderate (6 h)
Mask	Poly silicon	Stainless steel shadow mask	No
Etched surface for post bonding process	Very good	Poor due to particles	Very good

Table 1.
Advantage and disadvantage of the micromachining techniques in view of 500 μm through glass machining process and post anodic bonding of SOI wafer for differential pressure sensors.

Mask numbers	Masking layer	Etching depth (μm)	Etch selectivity
M1	Az4562	20	Poor
M2	a-Si/Cr/Au	100	Good
M3	0.5 μm Molybdenum	150	Good
M4	Polycrystalline silicon	350	Excellent

Table 2.
Summary of glass machining using different etch mask in 32% hydrofluoric acid etchant.

process complexities such as isotropic etch profile and failure in etch mask for longer etch duration. **Table 2** summarizes the etch trials for glass using 32% HF and 3.2% HCl as an etchant and different etch masks. HF was used to etch glass, while the small quantity of HCl increases the etch byproduct dissolution and enhances surface quality [17].

The investigation results show that M1 (Az4562 photoresist) can etch glass up to 20 μm. However, this etch mask is more prone to failure due to its poor adhesion and damaged by high HF concentration. Further, we have tried three different etch masks such as M2 (stack of a-Si/Cr/Au, 300/20/100 nm thickness), M3 (0.5 μm thick molybdenum) and M4 (1 μm thick poly-Si) as etch masks. Glass etched depths of 100, 150 and 350 μm were achieved successfully for M2, M3 and M4 respectively as tabulated in **Table 2**. As per this table, beyond the mentioned etch depths, the glass machining process was unsatisfactory due to etch mask failure. It was observed that the continued etching beyond a point resulted in pinhole generation on the etch mask due to high HF concentration. Though HCl was used, the etch byproducts redeposited on the etched surface and inhibited the glass etching, resulting in high surface roughness due to longer etch time. **Figure 8** shows the high surface roughness due to pinhole generation on poly-Si after 2 h of etching in HF. **Table 2** shows masks M2, M3, and M4 which can be used to etch glass for a depth of 50 μm (Photomask 6). It must be noted that while performing glass etch on one side, the other side has to be protected using the etch mask. In this aspect, two-sided deposition of poly-Si in LPCVD is advantageous over other etch masks such as M2 and M3. This eliminates the additional etch mask deposition step on the rear side of the glass wafer.

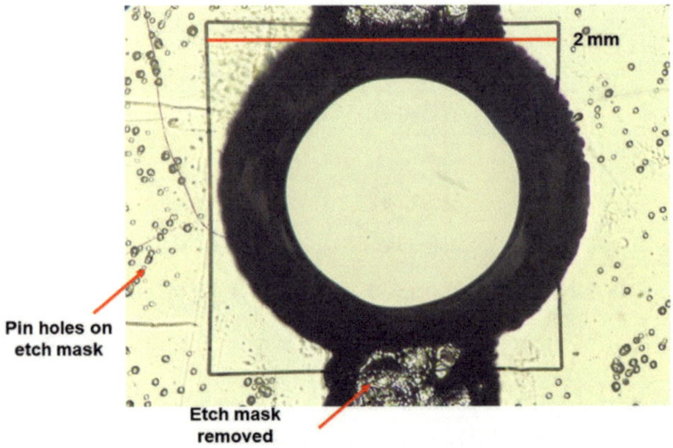

Figure 8.
Optical image of glass through etch by wet etch using HF solution.

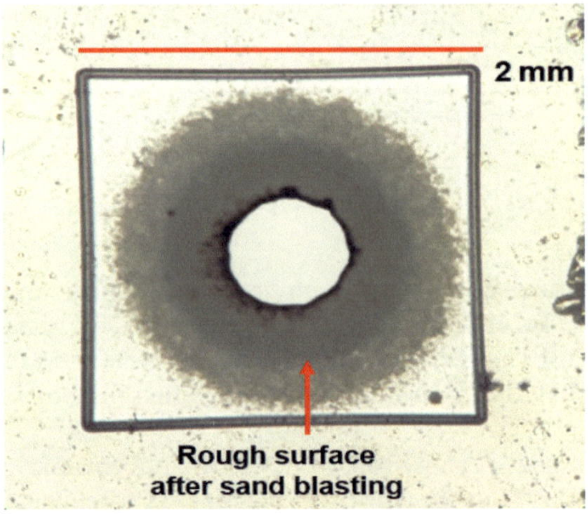

Figure 9.
Optical image of glass through etch by sand blasting.

For glass etching up to 350 μm depth, wet etch is preferred, however for the glass through etch sand blasting method is considered. Sand blasting is a physical process technique where a high-speed particle jet colloids with the glass surface resulting in mechanical erosion through a shadow mask of 0.8 mm thick stainless steel sheet. This shadow mask was aligned to the glass wafer so that circular pattern aligns at the centre of the square pattern. During the sand blasting process, abrasive sand particles of 50 μm size from the nozzle with a high-velocity stream impinges onto the glass substrate, causing glass etching. The process is simple, easy and drills 500 μm thick glass within 30 min. The biggest challenge involved in sand blasting is the glass wafer cleaning for the anodic bonding process. Also, as soon as the 500 μm etch is completed, the sand particles exit through the hole and etch the glass on the other side of the wafer up to 500 μm circumference (**Figure 9**). Thus, creating high surface roughness and particle contamination across the wafer.

Though the sandblasting process is successful in through glass etch, we have explored the ECDM technique in view of surface quality for anodic bonding.

5.2 Glass pattern by electro chemical discharge machining (ECDM)

Electrochemical discharge machining (ECDM) also called electrochemical arc machining (ECAM) is a combination of electric discharge machining (EDM) and electrochemical machining (ECM) techniques. Both EDM and ECM methods are limited only to conducting materials. This drawback was overcome by hybrid.

ECDM process, which can remove materials from both conductive (process is called as ECAM) and non-conductive (process is called as ECDM) materials [18]. This technique is undergoing lot of research and has much scope in commercializing it. The experimental setup is as shown in **Figures 10** and **11**. KOH solution was used as an electrolyte. The glass machining by ECDM undergoes thermal and chemical etching. The following reactions occur at the anode Eq. (1) and cathode Eq. (2) upon the applied voltage.

At the anode,

$$4OH^{-} \rightarrow 2H_2O + O_2 + 4e^{-} \quad (1)$$

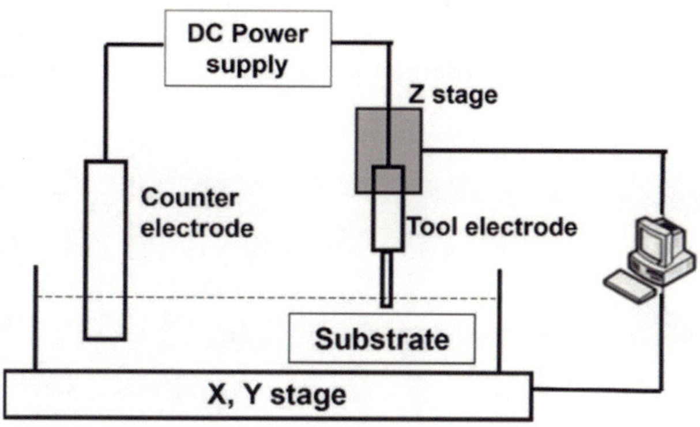

Figure 10.
Electro chemical discharge machining (ECDM) tool- schematic.

Figure 11.
Electro chemical discharge machining (ECDM) tools at Meukron technologies private limited, Bangalore, India.

At the cathode,

$$2H_2O + 2e^- \rightarrow 2OH^- + H_2 \qquad (2)$$

At a critical voltage, the spark is produced by the electrical discharge in the generated hydrogen gas layer. Due to the thermal energy of the spark and chemical reaction the material in contact will be eroded.

Generally, the glass etch profile by ECDM machining was improved by varying the experimental parameters such as type of electrolyte and its concentration, voltage, pulse on/off-time ratio, and feed rate [19]. In the current study, ECDM process was optimized by varying the experimental parameters to get a crack free etch profile with a slow etch rate of 1.4 μm/min. Importantly, the surface roughness after ECDM etching of the glass wafer which was exposed to the electrolytic solution was measured using atomic force microscopy. It was found that the surface roughness was <10 nm, favorable to anodic bonding process. The micromachined glass through etch is shown in **Figure 12**.

Finally, in comparison to wet etch and sand blasting process, the ECDM resulted in best of etch profile and smooth surface. Hence, the ECDM process was used to through etching of the glass wafer using photomask 7 for MEMS differential pressure sensors fabrication. The resulted glass substrate was bonded anodically to the SOI wafer (Section 4.1) **Figure 13** represents a cross sectional view of sensor die after bonding.

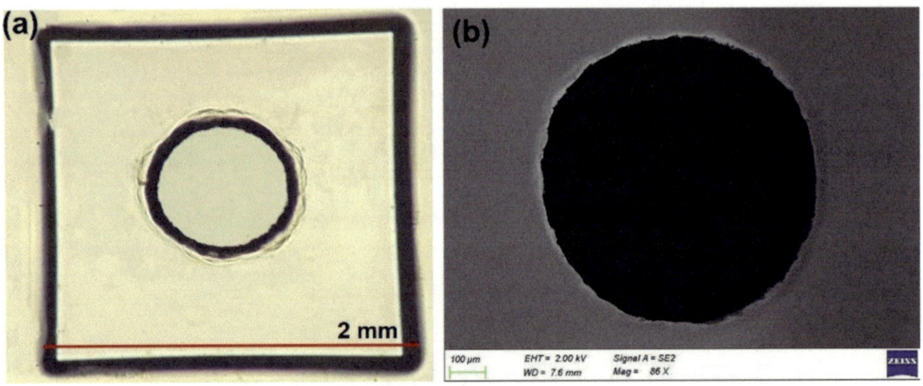

Figure 12.
Images of glass through etch by electro chemical discharge machining (a) Optical and (b) SEM images.

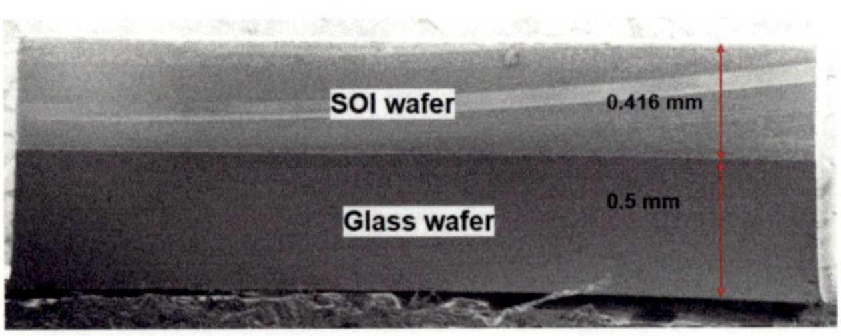

Figure 13.
Cross section SEM image of the SOI-glass bonding interface.

The fabricated wafer was diced, the individual sensor die (**Figure** 7) was separated, mounted on the header (also called glass to metal seal for hermeticity and high electrical insulation), proceeded further for wire bonding and testing. The detailed post fabrication steps are briefed in sections 6 and 7.

6. Packaging steps

Figure 14 shows the sequence of packaging steps. Fabricated headers (**Figure 15a** and **b**) are visually inspected using the stereo microscope for defect if any. Further the electrical insulation resistance test was carried out by applying 50 Volts between the body and the individual pins, the insulation should be greater than 100 MΩ. To test

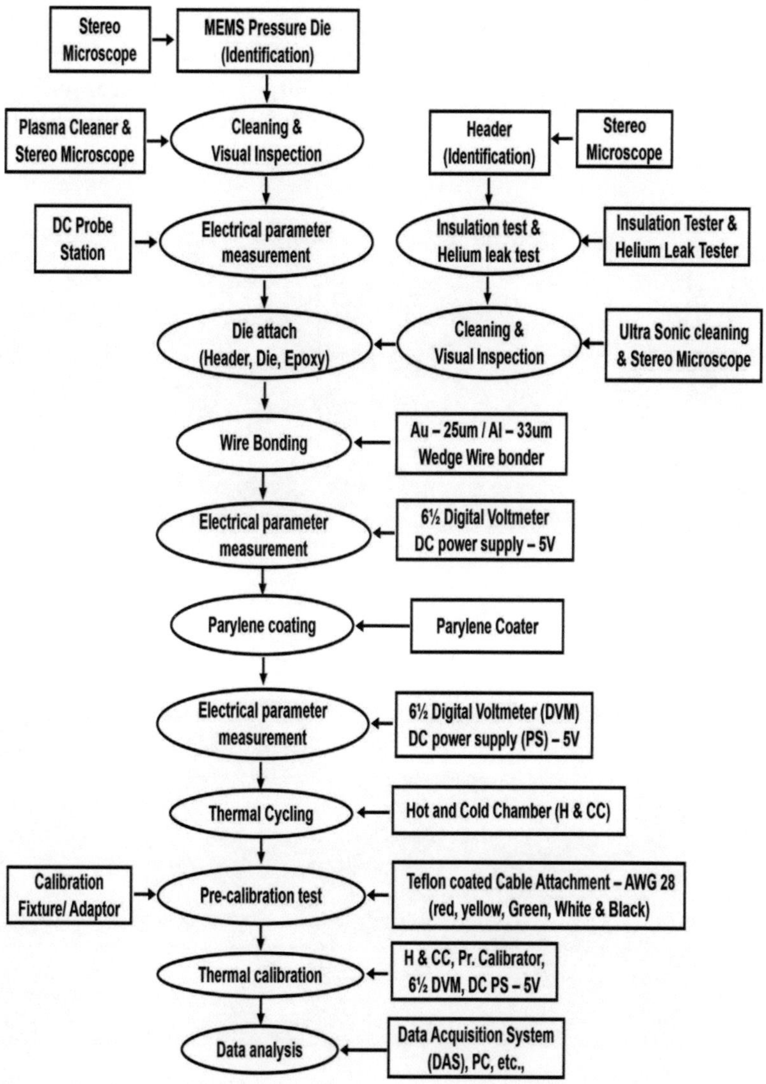

Figure 14.
Packaging process step flow chart.

the hermeticity, helium leak test was conducted to verify the leak rate between the header and pins, which must be minimum of 10^{-8} scc/s. The cleanliness of the headers was confirmed by visual inspection.

Identified MEMS pressure sensor die (**Figure 16b**) was cleaned using plasma cleaner for 10 min. Using DC probe station (**Figure 16a**), individual piezo resistances R1, R2, R3 and R4 were measured. **Figure 17a** and **b** shows the sensor die mounted on header with suitable epoxy and cured at room temperature for 24 h.

Figure 18 shows the wire bonding (33 μm Al) in open Wheatstone bridge configuration (**Figure 1b**). Pull strength of the bonded wires were measured which was 9 ± 1 grams (**Figure 18b**). Measurement of individual resistances was carried out using a 6½ digit voltmeter. Imbalance (offset voltage) of the closed Wheatstone bridge was measured using 5 V DC power supply. To protect the device from environmental dust, dirt, and humidity, parylene C was coated using the Parylene Coater, further individual resistance was measured using 6½ digit voltmeter. The offset of the Wheatstone bridge was measured with 5 Volts DC excitation under ambient conditions. For stress relieving process, thermal cycling of the wired die was carried out using the hot and cold chamber (H &CC). To measure the output of the sensor with pressure (0–140 mbar), a pneumatic calibrator was used (Fluke make) with the 5Volts excitation (**Figure 19**).

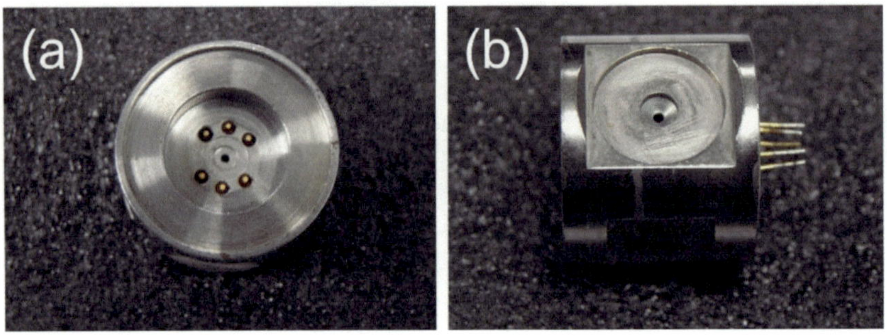

Figure 15.
(a) Header top view and (b) header side view.

Figure 16.
(a) DC probe station setup and (b) probing on MEMS pressure sensor die.

Figure 17.
(a) Die mounted on the header and (b) closure view of MEMS pressure sensor die attached on the header.

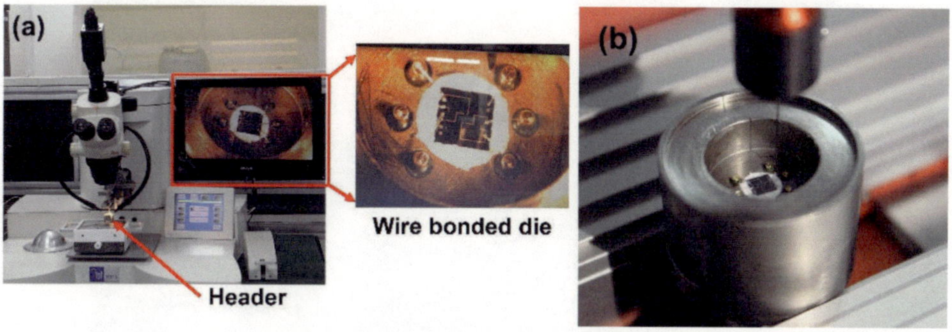

Figure 18.
(a) Wire bonding tool setup. (b) Pull strength testing of bonded wires.

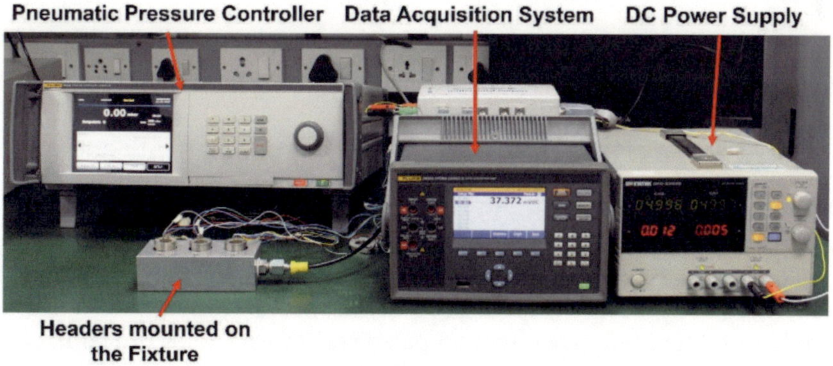

Figure 19.
Pneumatic pressure calibrator setup.

Initial pressure calibration in five ascending and five descending steps of equal pressure intervals was carried out to find out the zero-offset value, full scale output, sensitivity, nonlinearity, and hysteresis (**Table 3**). Further, varying humidity tests at constant temperature (22°C) and direct environmental exposure were carried out to ascertain drift or shift of the sensor output due to humidity effect (Section 7) using Data Acquisition System (DAS).

Sensor dice	With glass bond	With Cr/Au
Zero offset voltage (mV)	35.99	36.01
Sensitivity (mV/mbar)	0.73	0.78
Span (mV)	102.47	108.91
MAX NL + HYS (% F.S.O)	0.42	0.52

Table 3.
Consolidated data on pressure calibration for dice with glass bond and Cr/Au layer.

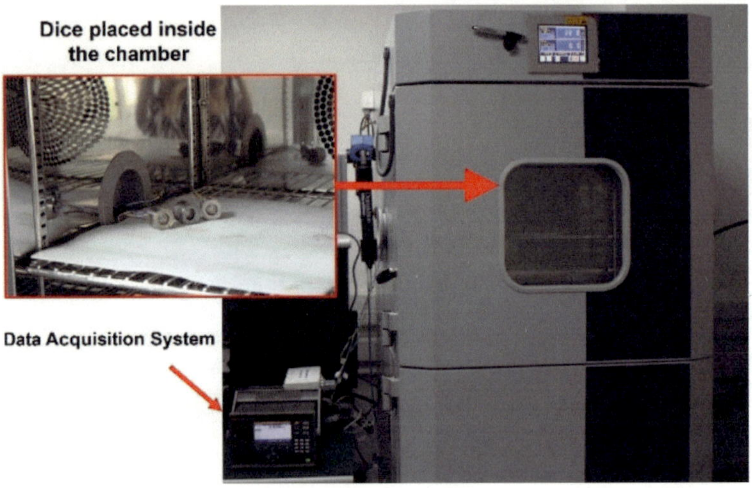

Figure 20.
Humidity experiment setup: bare die, dice with glass bonding and Cr/Au passivation placed inside the chamber.

7. Pressure calibration and effect of humidity on piezo resistors

Humidity exposure experiments were conducted using bare die and two more sensor dice one with glass bonding and another with Cr/Au passivation on the rear side. Pressure calibration (**Figure 19**) was conducted and measured zero offset values, and output voltages for five ascending and descending pressure of 28, 56, 84, 112 and 140 mbar with an excitation voltage of 5 V DC at 25°C (**Table 3**). After recording the pressure response data, the sensors were subjected to study the humidity and temperature effect on piezo resistance values in a controlled environment as well as direct exposure to the atmosphere. The humidity effect in a controlled environment (**Figure 20**) was performed in a closed chamber with pre-defined humidity values of 30, 60 and 90% Rh with 2 h of exposure time. From the experimental results, the percentage change in piezo resistance values for the bare die, glass bonded die and die having Cr/Au passivation was found to be 0.106, 0.035 and 0.034% respectively (**Figure 21**).

Further, a similar experiment was conducted by exposing the sensor dice directly to the atmosphere by placing the sensors outside the window for 2 days (**Figure 22**). The resistance values change due to the environmental conditions were recorded

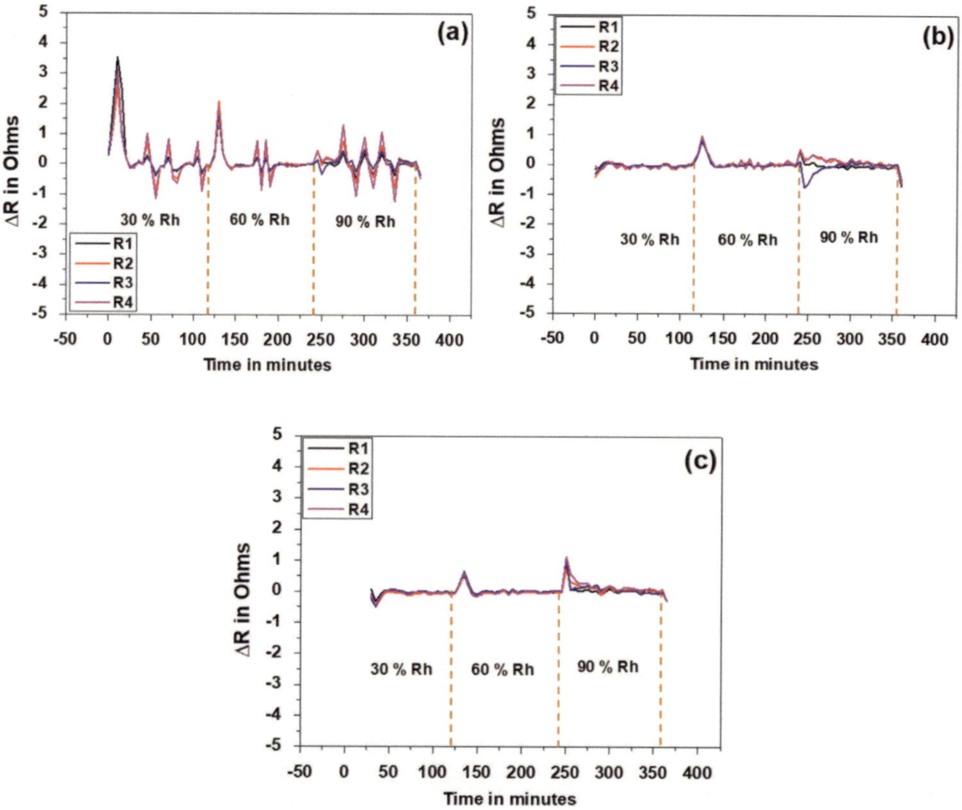

Figure 21.
Effect of humidity on piezo resistance for the sensors placed inside a humidity chamber (a) bare die (0.106% resistance change) (b) with glass bond (0.035% resistance change) and (c) with Cr/Au passivation layer (0.034% resistance change).

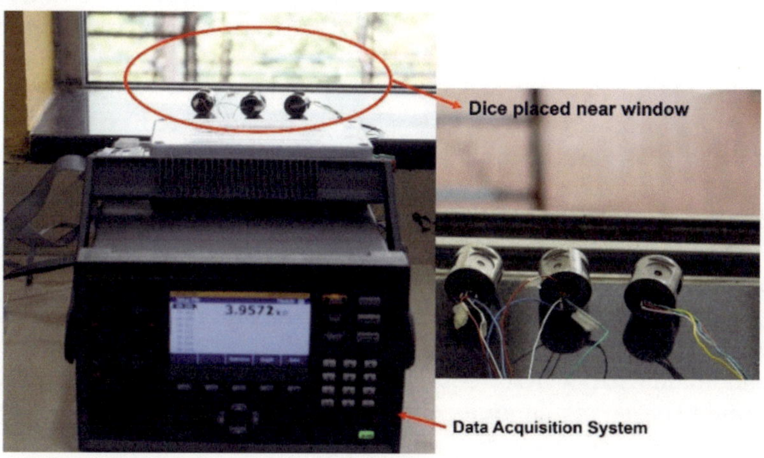

Figure 22.
Humidity experiment set up: dice with & without glass bonding and Cr/Au passivation exposed to direct environment by placing them near the window.

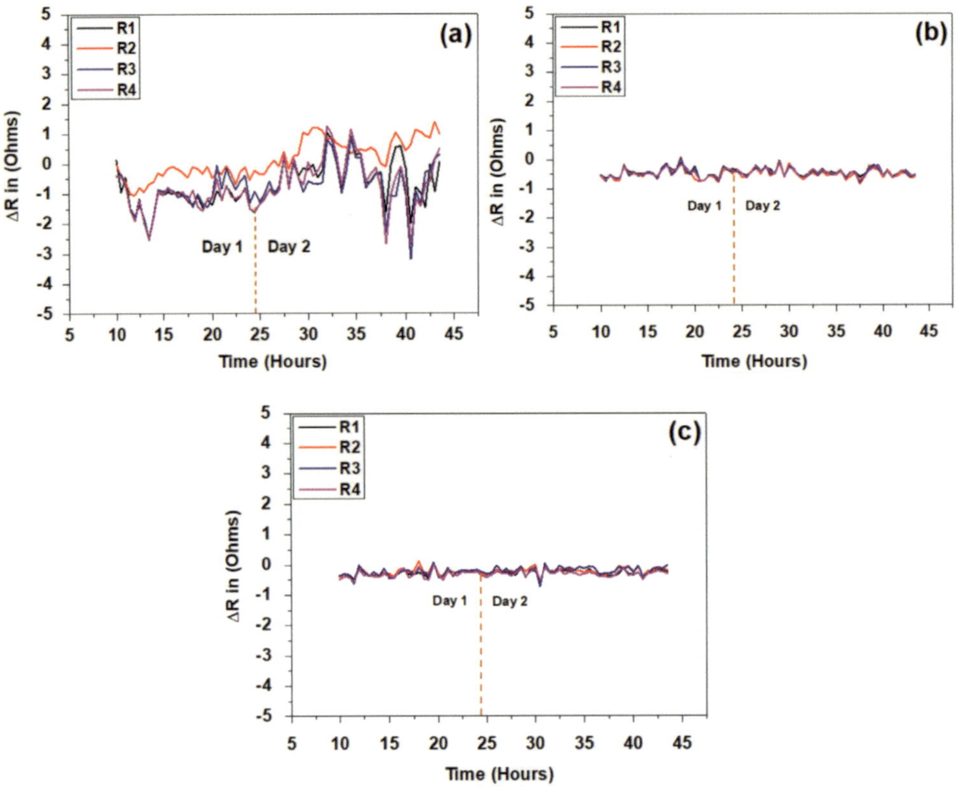

Figure 23.
Effect of humidity on piezo resistance for the sensors exposed directly to the atmosphere (a) without glass bond (0.092%resistance change) and (b) with glass bond (0.026% resistance change) and (c) with Cr/Au passivation layer (0.022% resistance change).

using DAS. **Figure 23** shows the percentage change in piezo resistance value for the sensors without glass bond, with glass bond & Cr/Au layer which were found to be 0.092, 0.026 and 0.022% respectively.

As compared to the bare die, piezo resistance change due to humidity is nearly four times better for other two dice. Cr/Au layered die found to be slightly better than anodic bonded die. The glass bonding to the sensor die provides thermal stability and protection from the environmental dust particles. The Cr/Au passivation layer protects the die from the absorption/adsorption of moisture on the sensors surface. This was confirmed by measuring the sensor's zero offset values and output response to the applied pressure after subjecting it to extreme humidity tests (**Table 3**).

Table 3 shows zero offset values, sensitivity, span, and maximum non linearity and hysteresis. The zero offset values for Cr/Au and glass bonded matches closely (0.05%) and nonlinearity + hysteresis 0.1%. Whereas sensitivity 0.05 mVolt/mbar. Bare die found to be having humidity effect and hence zero offset voltage variation.

8. Conclusions

The present chapter reports the improved MEMS differential pressure sensor output characteristics by incorporating new modifications in the fabrication stage.

The environmental humidity and temperature variation directly affect the offset of the sensor. The addition of glass bonding at the rear side of the pressure sensor die enhances thermal stability and reduces the sensor's offset drift. The Cr/Au passivation layer minimizes and almost eliminates the absorption/adsorption of moisture content on the sensor surface, which results in reduced offset drift. The critical challenges involved in glass machining by wet etching, sandblasting, and electrochemical discharge machining (ECDM) were discussed in view of glass bonding process. It was found that the fabrication of differential MEMS pressure sensor complexities were minimized in case of ECDM technique in view of glass machining and anodic bonding processes.

Acknowledgements

We acknowledge funding support from MHRD (Ministry of Human Resource Development), MeitY and DST (Department of Science and Technology) Nano Mission, India through NNetRA. The authors would also like to acknowledge the National Nano Fabrication Centre (NNFC), the Micro and Nano Characterization Facility (MNCF) and Advanced systems and packaging lab of the Centre for Nano Science and Engineering, Indian Institute of Science for providing access to the fabrication, characterization and testing facilities. The authors also acknowledge the Meukron Technologies Private Limited, Bangalore for ECDM of glass.

Conflict of interest

The authors declare no conflict of interest.

Author details

Bagur Rudrappa Deepu*, Belthangady Pavithra, Mangalore Manjunatha Nayak and Kunchinadka Narayana Bhat
Centre for Nano Science and Engineering, Indian Institute of Science, Bangalore, India

*Address all correspondence to: deepabr31@gmail.com

References

[1] Bhat KN, Nayak MM. MEMS pressure sensors: An overview of challenges in technology and packaging. ISSS Journal. 2012;**1**:1-10

[2] Geyling FT, Forst JJ. Semiconductor strain transducers. Bell System Technical Journal. 1960;**39**:705-731. DOI: 10.1002/j.1538-7305.1960.tb03939.x

[3] Smith CS. Piezoresistance effect in germanium and silicon. Physical Review. 1954;**94**:42-49. DOI: 10.1103/PhysRev.94.42

[4] Tran AV, Zhang X, Zhu B. Effects of temperature and residual stresses on the output characteristics of a piezoresistive pressure sensor. IEEE Access. 2019;**7**:27668-27676. DOI: 10.1109/ACCESS.2019.2901846

[5] Cao G, Wang X, Xu Y, Liu S. A micromachined piezoresistive pressure sensor with a shield layer. Sensors (Switzerland). 2016;**16**:1-13. DOI: 10.3390/s16081286

[6] French PJ, Muro H, Shinohara T, Nojiri H, Kaneko H. SOI pressure sensor. Transducers. 1991;**35**:181-184. DOI: 10.1016/0924-4247(92)87003-Y

[7] Guo S, Eriksen H, Childress K, Fink A, Hoffman M. High temperature smart-cut SOI pressure sensor. Sensors Actuators, Applied Physics. 2009;**154**:255-260. DOI: 10.1016/j.sna.2009.03.011

[8] Gaska R, Shur MS, Bykhovski AD, Yang JW, Khan MA, Kaminski VV, et al. Piezoresistive effect in metal-semiconductor-metal structures on p-type GaN. Applied Physics Letters. 2000;**76**:3956-3958. DOI: 10.1063/1.126833

[9] Wejrzanowski T, Tymicki E, Plocinski T, Bucki JJ, Tan TL. Design of SiC-doped piezoresistive pressure sensor for high-temperature applications. Sensors. 2021;**21**:1-16. DOI: 10.3390/s21186066

[10] Belwanshi V, Topkar A. Quantitative analysis of MEMS piezoresistive pressure sensors based on wide band gap materials. IETE Journal of Research. 2022;**68**:667-677. DOI: 03772063.2019.1620641

[11] Mosser V, Suski J, Goss J, Obermeier E. Piezoresistive pressure sensors based on polycrystalline silicon. Sensors and Actuators, A: Physical. 1991;**28**:113-132. DOI: 10.1016/0924-4247(91)85020-O

[12] Peng CT, Lin JC, Te LC, Chiang KN. Performance and package effect of a novel piezoresistive pressure sensor fabricated by front-side etching technology. Sensors and Actuators, A: Physical. 2005;**119**:28-37. DOI: 10.1016/j.sna.2004.08.013

[13] Lisec T, Kreutzer M, Wagner B. Surface micromachined piezoresistive pressure sensors with step-type bent and flat membrane structures. IEEE Transactions on Electron Devices. 1996;**43**:1547-1552. DOI: 10.1109/16.535348

[14] Tummala RR. Fundamentals of Microsystems Packaging. Singapore: McGraw-Hill Education; 2001. pp. 606-607

[15] Bhat KN et al. Design, development, fabrication, packaging, and testing of MEMS pressure sensors for aerospace applications. In: Vinoy K, Ananthasuresh G,

Pratap R, Krupanidhi S, editors. Micro and Smart Devices and Systems. New Delhi: Springer; 2014. pp. 3-17. DOI: 10.1007/978-81-322-1913-2_1

[16] Kumar SS, Pant BD. Design principles and considerations for the 'ideal' silicon piezoresistive pressure sensor: A focused review. Microsystem Technologies. 2014;**20**:1213-1247. DOI: 10.1007/s00542-014-2215-7

[17] Iliescu C, Jing J, Tay FEH, Miao J, Sun T. Characterization of masking layers for deep wet etching of glass in an improved HF/HCl solution. Surface and Coatings Technology. 2005;**198**:314-318. DOI: 10.1016/j.surfcoat.2004.10.094

[18] Basak I, Ghosh A. Mechanism of spark generation during electrochemical discharge machining: A theoretical model and experimental verification. Journal of Materials Processing Technology. 1996;**62**:46-53. DOI: 10.1016/0924-0136(95)02202-3

[19] Cao XD, Kim BH, Chu CN. Hybrid micromachining of glass using ECDM and micro grinding. International Journal of Precision Engineering and Manufacturing. 2013;**14**:5-10. DOI: 10.1007/s12541-013-0001-6

Chapter 4

Force Measuring System with Zero-Compliance Suspension

Takeshi Mizuno, Masaya Takasaki and Yuji Ishino

Abstract

The principle and classification of force measuring system with zero-compliance suspension are described. A zero-compliance suspension consists of series-connected suspensions, one of which operates to cancel the variation of length of the other suspension. As a result, the total length of the connected suspensions is invariant even when force acts between the ends. In the force measuring system, one end is fixed to the base and force to be measured is applied to the other end (point of application). The force is estimated from the displacement of the connection point of the suspensions that is to be detected (detection point). This principle of measuring force is explained in a basic model. Measuring systems are classified according to the location of active element, the actuator of the active element, and the location of positive-stiffness element. Several examples of measurement system are described. Then, control methods of achieving zero-compliance are explained for an apparatus using double series magnetic suspension. Finally, measurement results by the developed apparatus are shown. It is confirmed experimentally that the displacement of the detection point is proportional to external force acting on the point of application and high-resolution measurement is possible by the developed system.

Keywords: force measurement, zero-compliance, infinite stiffness, servomechanism, magnetic suspension, MEMS technology

1. Introduction

Force is one of the most fundamental physical quantities. Accurate force measurement is required in a lot of engineering systems [1]. A variety of force measurement methods have been proposed and in practical use. They are classified in a number of different ways. One classification is according to the system structure: open-loop system or closed-loop system. The former corresponds to the deflection method [2, 3], while the latter corresponds to the null method [2–4].

Most of the developed devices are of the former type [5–7]. One of the most popular devices is load cell [1]. In devices of this type, the stiffness of the mechanical conversion part is made lower to achieve higher resolution. However, such low-stiffness mechanism has several drawbacks. One of them is the deflection of the point of application of force by the measurand (force). It may cause measurement error, because the distance between the force source and the point of application changes in measuring. Another is the limited bandwidth of measurement.

The first drawback can be overcome by the null method (second category). However, high-gain feedback is implemented to ensure against phase lag in measurement [8–10]. It often makes the control signal noisy, which may result in significant degradation of measurement resolution, because the force is estimated from the control signal.

Force measurement using zero-compliance mechanism has been proposed to overcome the above-mentioned problems [11]. The original concept of zero-compliance mechanism is just a series-connected two springs, one of which has positive stiffness and the other has negative stiffness, whose amplitude is equal to that of the former. In this condition, the total stiffness becomes infinite. It enables the point of application of force to keep the original position even when force acts on it like in the null method, while the connection point of the two springs (detection point) deflects in proportional to the applied force like in the deflection method.

In this chapter, the principle of measuring force is explained in a basic model. Then, measuring systems are classified according to the location of active element, the actuator of the active element, and the location of positive-stiffness element. Several examples of measurement system are briefly introduced. One of them is micromachined. It will suit microforce measurement because of its higher resolution. Then, control methods of achieving zero-compliance are explained in an apparatus using double series magnetic suspension. Finally, measurement results by the apparatus are shown to demonstrate the effectiveness of the proposed measuring system.

2. Principles of measurement

See **Figure 1**.

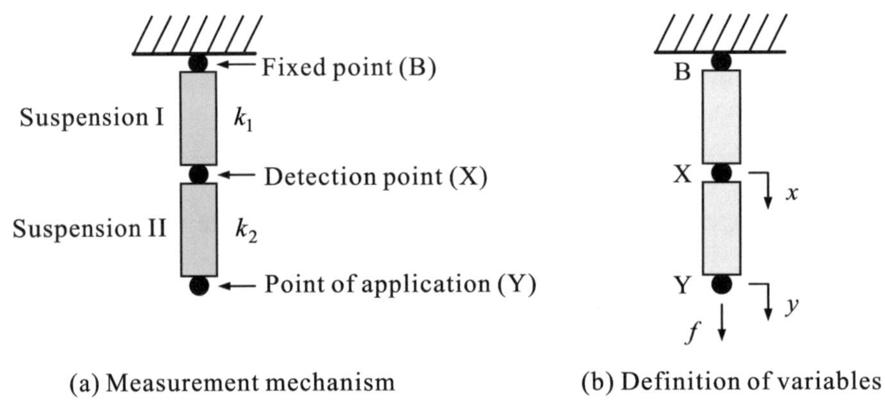

Figure 1.
Schematic drawings of measurement mechanism.

2.1 Zero-compliance mechanism

The principle of force measurement with zero-compliance mechanism is illustrated in **Figure 1**. Suspension I and Suspension II are connected in series at the point X. This point becomes the detection point. Force to be measured is applied to the other end of Suspension II (Y). This point is called the point of application. The stiffness of the series-connected suspensions becomes

$$k_c = \frac{k_1 k_2}{k_1 + k_2}, \tag{1}$$

where k_c: stiffness of the series-connected suspensions, k_i: stiffness of each suspension ($i = 1, 2$). This equation indicates that the resultant stiffness becomes lower than the stiffness of each suspension when both suspensions have positive stiffness ($k_i > 0$). Nevertheless, the situation dramatically changes if one of the suspensions is allowed to have negative stiffness. Especially, when

$$k_1 = -k_2 \tag{2}$$

is satisfied, the stiffness of the connected suspensions becomes infinite as

$$|k_c| = \infty. \tag{3}$$

It means that the point of application is invariant (no deflection), which is just a characteristic of measurement by the null method. Meanwhile, the detection point moves from the original position in proportion to the applied force as

$$x = \frac{f}{k_1} = -\frac{f}{k_2}, \tag{4}$$

where x: displacement of detection point and f: force acting on the point of application (**Figure 1b**). Therefore, the force can be estimated from the displacement of the detection point, which characterizes measurement by the deflection method. Eq. (4) indicates that each suspension should have low stiffness for high resolution in measurement.

2.2 Classification of mechanisms

In explaining the principle of operation in the previous section, pure springs with positive and negative stiffness were assumed. However, to achieve the zero-compliance characteristic by a stable system, active control is necessary [12]. It indicates that at least one of the suspensions must be an active element equipped with an actuator. Hence, the measurement systems can be classified according to [13, 14]

- location of active element,
- actuator used in active element,
- location of positive (or negative) stiffness element.

As to the first criterion of classification (location of active element), three ways are possible:

I. Suspension I,

II. Suspension II,

III. Suspension I and Suspension II.

Simply to achieve zero-compliance, (I) or (II) is sufficient. It is advantageous in reducing component count and cost. The third may be adopted when a more flexible system is required, for example, both stiffness are adjustable.

As to the second criterion of classification (actuator), the followings are candidates:

(M) Electromagnet,
(V) Voice coil motor,
(E) Electrostatic actuator,
(P) Piezoelectric actuator.

It is to be mentioned that any other linear actuator is applicable.

As to the third criterion of classification (location of positive-stiffness element),

(U) Suspension I,
(L) Suspension II.

Figure 2 shows how the mechanism operates in each case. In the upper location (U), the detection point moves in the same direction as the applied force. In the lower location (L), it moves in the direction opposite to the applied force.

3. Examples of mechanism and measuring apparatus

A simple calculation indicates 24 combinations for the realization of zero-compliance mechanism. The selection should be done according to target force (gravitational, magnetic, electrostatic, atomic, *etc.*), the magnitude/order of force (mega to pico-Newton), the size of instrument (macro- to micro-scale), and environment (clean, chemical, vacuum, operation temperature, etc.). The optimization is still under investigation.

Among the possible combinations, the first developed apparatus was a combination of (I)-(M)-(U) [15], whose system configuration and picture are shown in

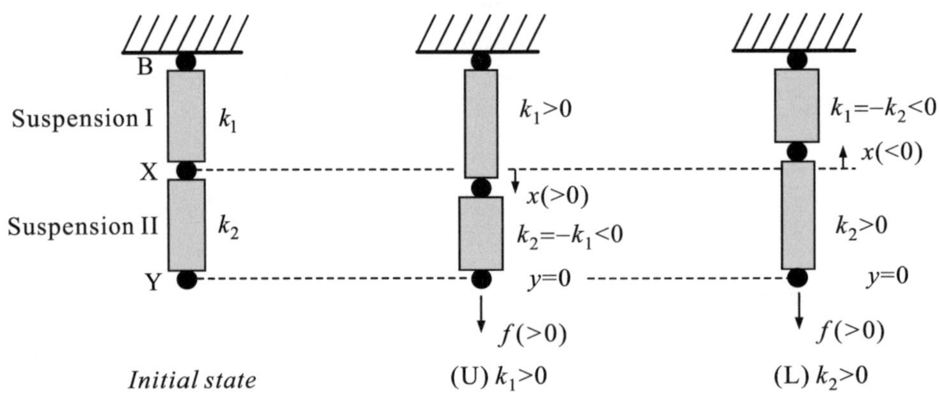

Figure 2.
Operation to applied force.

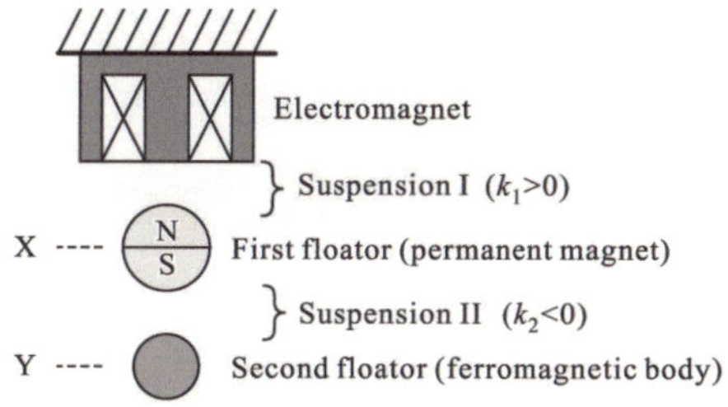

Figure 3.
Measurement system using double series magnetic suspension.

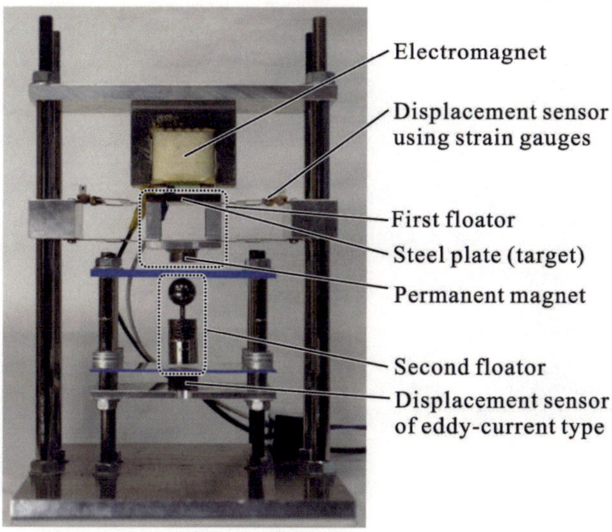

Figure 4.
Picture of fabricated apparatus using double series magnetic suspension.

Figures 3 and **4**. It was based on double series magnetic suspension. In this suspension system, two floators (objects to be suspended) are controlled with a single electromagnet. The force of the electromagnet directly acts on the first floator equipped with a permanent magnet at the bottom. It corresponds to the detection point. The attractive force of the permanent magnet acts on the second floator made of ferromagnetic material. It corresponds to the point of application. The force increases as the gap between the floators decreases. Thus, it acts as a spring with negative stiffness. When a downward force acts on the point of application, the second floator is indirectly controlled to keep the original position by the electromagnet. Simultaneously, the detection point deflects downward for the gap between the two floators to decrease, and the increased magnetic force cancels the force applied to the point of application to achieve zero-compliance state.

The second developed apparatus was a combination of (I)-(V)-(L) [16], whose system configuration is shown in **Figure 5**. In this configuration, Suspension II has positive stiffness and suspends the point of application from the detection point. The motion of the detection point is controlled by a VCM. **Figure 6** shows a picture of the fabricated apparatus. It uses a one-piece parallel spring mechanism where a VCM is installed with as an actuator as shown in **Figure 7**. Such structure is effective for

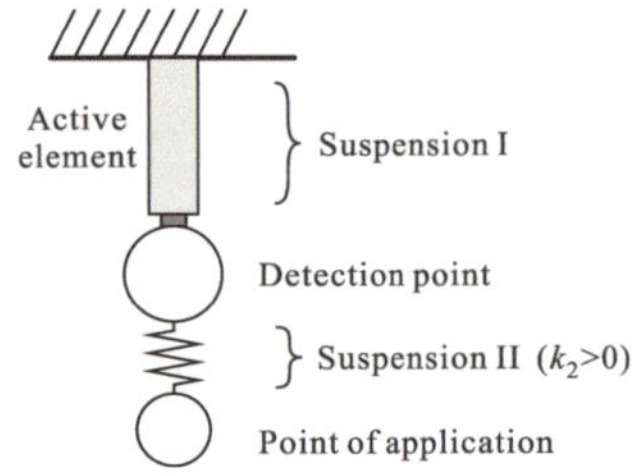

Figure 5.
System using positive spring and voice coil motor.

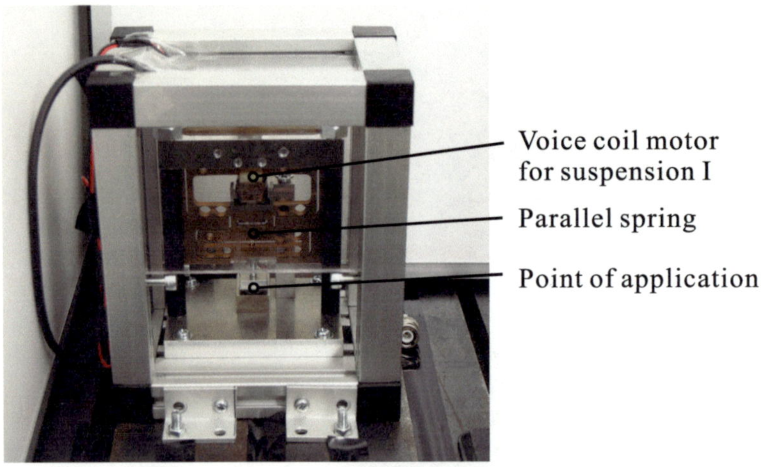

Figure 6.
Picture of fabricated apparatus with one-piece parallel spring mechanism and voice coil motor.

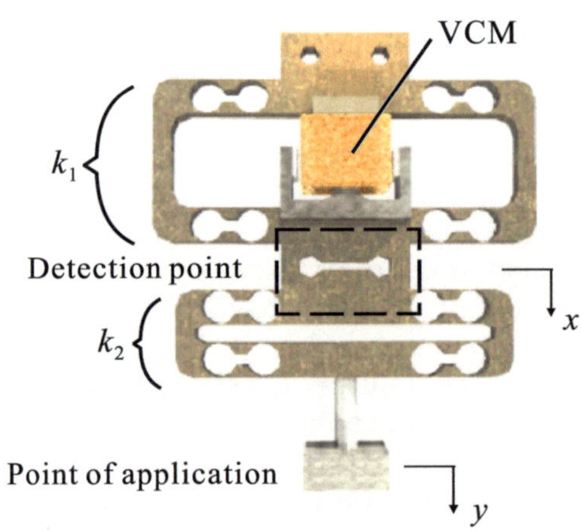

Figure 7.
Picture of parallel spring mechanism with VCM.

downsizing the measurement mechanism. The spring mechanism has sixteen circular hinges. It operates like a series-connected springs. The VCM is located between the top and the middle of the frame to make the stiffness of the upper spring negative.

The point of application is attached to the bottom of the frame. The motions of the point of application and the detection point are restricted solely to the vertical translation by the spring mechanism.

One of the more effective ways of downsizing the measurement mechanism is the application of MEMS technology in which electrostatic actuator is mostly used. An apparatus based on MEMS technology is a combination of (I)-(E)-(U), whose configuration is shown in **Figure 8**. A photo and a schematic drawing of the measuring device are shown in **Figure 9** [17]. Electrostatic actuator is used in Suspension I (active element), whose electrode is shown in **Figure 10**. The displacement sensors

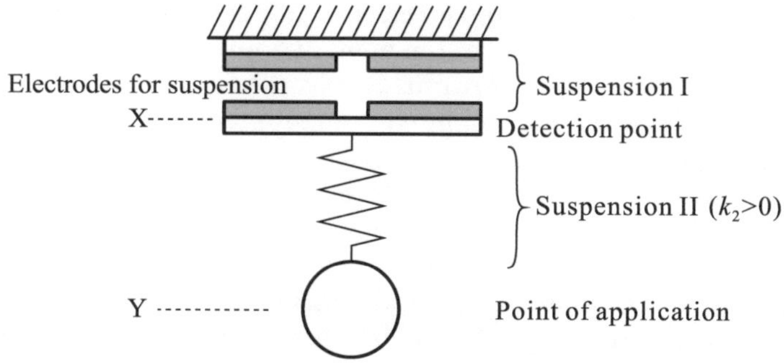

Figure 8.
System using positive spring and electrostatic actuator.

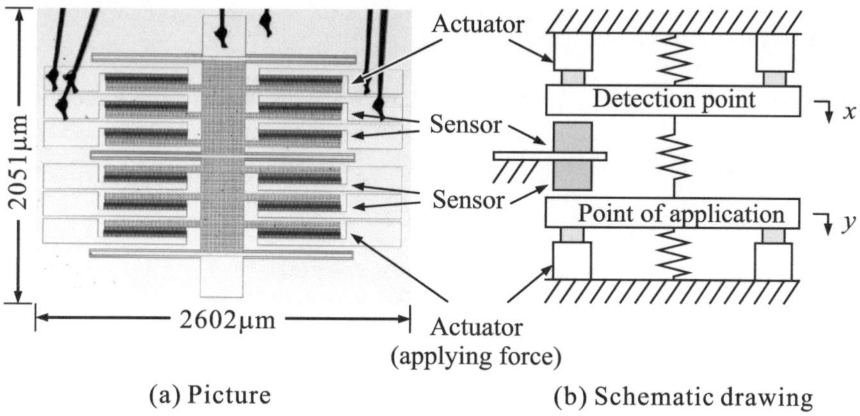

Figure 9.
Measurement device fabricated by MEMS technology. (a) Picture. (b) Schematic drawing.

Figure 10.
Enlarged picture of comb-shaped electrode.

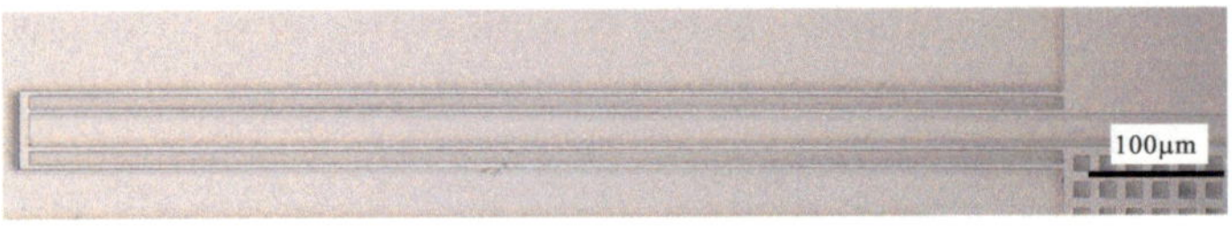

Figure 11.
Enlarged picture of leaf spring.

are capacitive and have same electrodes. Suspension II (positive-stiffness element) is also fabricated by MEMS technology as shown in **Figure 11**. To restrict the motion of the point of application to a single translational motion, the point of application is also suspended by this spring. In addition, an electrostatic actuator is also implemented to the point of application to give load to the point of application for testing the performances. Fine force measurement performances have been improved dramatically [17]. It demonstrates that the applicability of the proposed force measuring system will be widened by micromachining.

The unique features of the zero-compliance-based measuring system are advantageous in measuring fine force that is a function of the distance from the source of force, for example, magnetic or atomic force. The critical problem to industrial use is that the complexity and component count of system inevitably increase in comparison to conventional measuring system, mainly because an active element is necessary.

4. Control and measurement results

4.1 Mathematical model

Control methods of achieving zero-compliance are discussed in this section. The measurement system shown in **Figure 4** is targeted. The physical model is shown in **Figure 12**. The equations of motion are given by [13, 15]

$$m_d\ddot{x}(t) = -F_e + F_p + m_d g \tag{5}$$

$$m_p\ddot{y}(t) = -F_m + m_p g + f(t) \tag{6}$$

where x: displacement of the detection point, y: displacement of the point of application, m_d: mass of detection point, m_p: mass of the point of application, F_e:

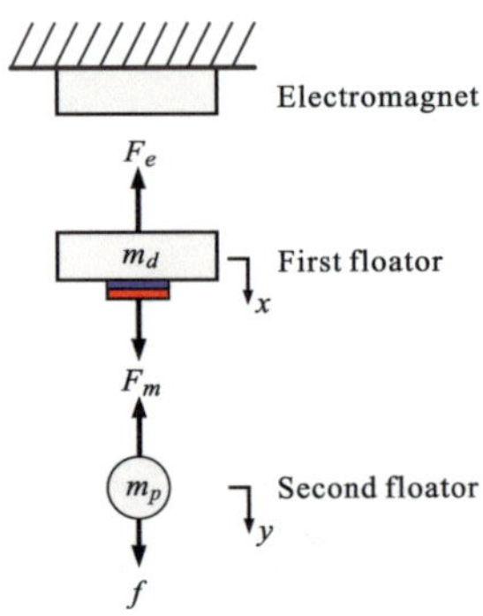

Figure 12.
Physical model of measurement system.

attractive force of the electromagnet F_m: attractive force of the permanent magnet, and f: force acting on the point of force. In the actual apparatus, the first floator is suspended by leaf spring to restrict the motion to a single-degree-of-freedom translational motion [15]. However, this effect is neglected here for simplification. It is to be noted that this effect has no influence on accuracy in static force measurement theoretically. In the neighborhood of the equilibrium states, the forces F_e and F_m are approximately given by

$$F_e = (m_d + m_p)g - k_i i + k_s x \tag{7}$$

$$F_m = m_p g + k_m(x - y) \tag{8}$$

where i: control current, k_i: coefficient of linearized characteristic equation of electromagnet (ratio of force to current), k_s: coefficient of linearized characteristic equation of the electromagnet (ratio of force to gap), k_m: coefficient of linearized characteristic equation of the permanent magnet (ratio of force to gap). The last coefficient k_m is equal to the absolute value of the negative stiffness of Suspension II. The substitution of Eq. (7) and Eq. (8) into Eq. (5) and Eq. (6) leads to the linearized equations of motion:

$$m_d \ddot{x}(t) = k_s x + k_m(x - y) - k_i i, \tag{9}$$

$$m_p \ddot{y}(t) = -k_m(x - y) + f(t), \tag{10}$$

Laplace transforming Eq. (9) and Eq. (10) with null initial values leads to

$$m_d s^2 X(s) = -(k_s + k_m)X(s) - k_m Y(s) - k_i I(s), \tag{11}$$
$$m_p s^2 Y(s) = k_m(X(s) - Y(s)) + F(s), \tag{12}$$

where capitals are Laplace-transformed variables.

4.2 Controller

To achieve zero-compliance, there are several applicable control methods, such as

i. Integral feedback of the displacement of the point of application;

ii. Displacement cancellation control;

iii. Stiffness control includes proportional control and zero-power control [13].

We focus on the method (i). When the control method (i) with state feedback is applied, the control input is represented by

$$I(s) = (p_d + p_v s)X(s) + \left(q_d + q_v s + \frac{q_I}{s}\right)Y(s), \tag{13}$$

where p_d and p_v: proportional and derivative feedback gains related to detection point, q_d, q_v, and q_I: proportional, derivative, and integral feedback gains related to point of application.

4.3 Analysis

Substituting Eq. (13) into Eqs. (11) and (12) and rearranging them gives

$$X(s) = -\frac{k_i q_v s^2 + (k_m + k_i q_d)s + k_i q_I}{\hat{t}_c(s)} F(s), \tag{14}$$

$$Y(s) = \frac{m_d s^2 + k_i p_v s + k_i p_d - (k_m + k_s)}{\hat{t}_c(s)} F(s), \tag{15}$$

where

$$\begin{aligned}\hat{t}_c(s) = m_d m_p s^5 + m_p k_i p_v s^4 + \left[m_p\{k_i p_d - (k_m + k_s)\} - m_d k_m\right]s^3 \\ -k_m k_i(p_v + q_v)s^2 - k_m\{k_i(p_d + q_d) - k_s\}s - k_m k_i q_I\end{aligned} \tag{16}$$

Assuming that the force acting on the point of force is constant as

$$F(s) = \frac{f_0}{s}, \tag{17}$$

and the closed-loop is stable, the steady-state displacements are given by

$$x(\infty) = \lim_{t\to\infty} x(t) = \lim_{s\to 0} sX(s) = \frac{f_0}{k_m} \tag{18}$$

$$y(\infty) = \lim_{t\to\infty} y(t) = \lim_{s\to 0} sY(s) = 0 \tag{19}$$

It is proved by Eq. (19) that the point of application is invariant in position (see **Figure 2**(U)). It is shown by Eq. (18) that the detection point moves in the same direction of the applied force; the displacement is larger as the coefficient k_m is smaller. From Eqs. (18) and (19), we get

$$y(\infty) - x(\infty) = -\frac{f_0}{k_m}. \tag{20}$$

It indicates that the distance between the detection point and the point of application decreases when the applied force is positive (downward) as if the stiffness between them is negative.

4.4 Measurement results

In the following figures, downward (positive) displacement is plotted to the downward direction for intuitive understanding. **Figure 13** shows the control current, the displacement of the point of application, and the displacement of the detection point when static force is added to the point of application [15]. The static force was produced by adding weights to the point of application one by one. One weight corresponds to 24.5 [mN]. After 10 weights were added to the floator, weights were taken off one by one. This result demonstrates that the point of application is almost invariant, while the detection point moves proportionally to the applied force. It is also found that there are slight differences between increasing and decreasing. It was caused by magnetic hysteresis. Meanwhile, the force can be estimated from the

DOI: http://dx.doi.org/10.5772/intechopen.114102

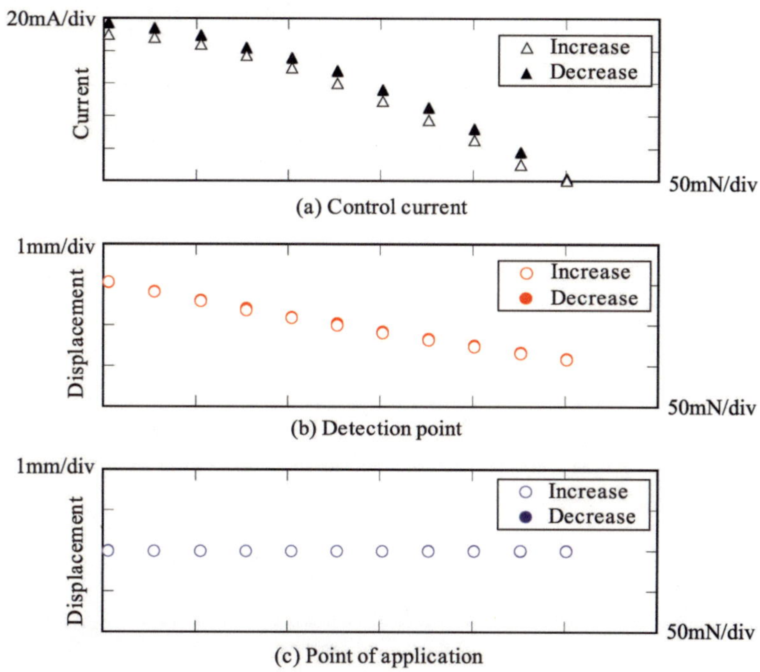

Figure 13.
Measurement results; added removed weight corresponds to 24.5 mN.

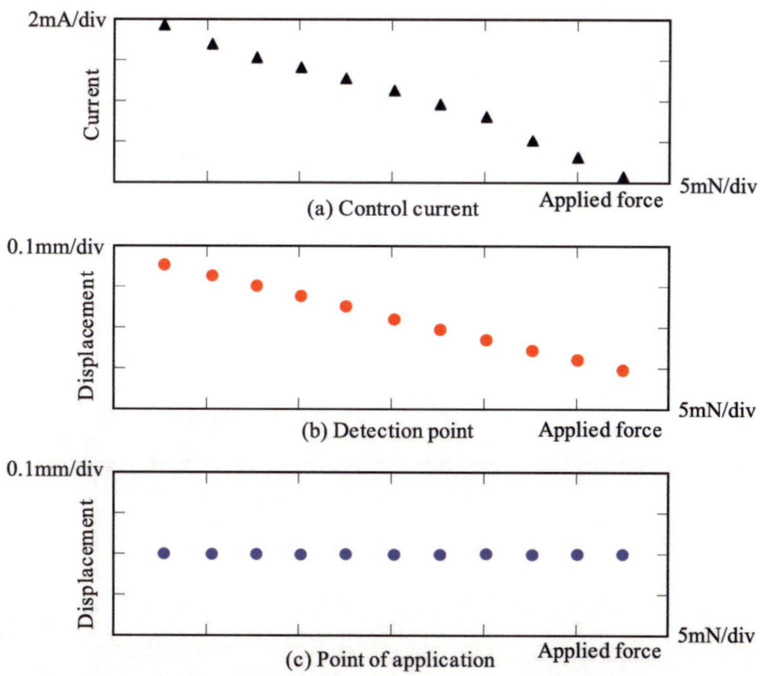

Figure 14.
Measurement results; added weight corresponds to 2.45 mN.

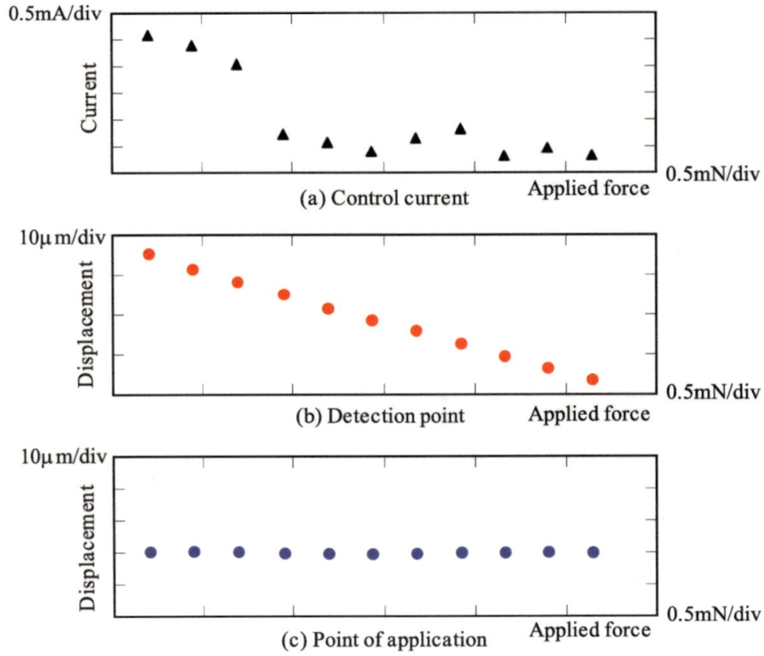

Figure 15.
Measurement results; added weight corresponds to 0.245 mN.

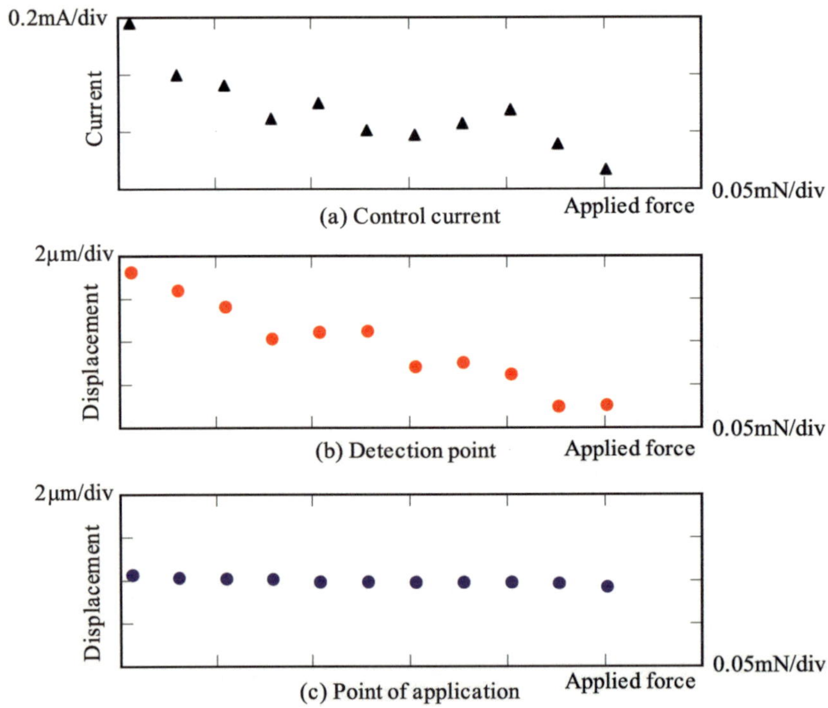

Figure 16.
Measurement results; added weight corresponds to 0.0245 mN.

control input (current) like by the conventional null method. Comparing it to the displacement of the detection point indicates that the linearity is worse.

To study the resolution of the measurement system, smaller weights were added to the point of application. **Figures 14–16** show the measurement results when one weight corresponds to 2.45 [mN] (**Figure 12**), 0.245 [mN] (**Figure 13**), and 0.0245 [mN] (**Figure 14**). **Figure 14** shows that both the control current and the displacement of the detection point vary proportionally according to the applied force; the linearity is again better in the latter. As shown by **Figure 15**, the displacement of the detection point is still proportional to the applied force (b), while such linearity is lost in the control current (a). This is due to noise including in the control signal. In **Figures 13–15**, the displacement of the detection point is proportional to the applied force. In contrast, such clear linear relation is not found in **Figure 16**. These results indicate that the resolution is 0.245-mN order in the developed apparatus. The mass of the floator with no weight is 69.2 [g]. It indicates that a change of 0.036% (the ratio of 0.025 to 69.2 [g]) in force can be detected by the proposed method.

5. Conclusions

The principle of force measurement using zero-compliance suspension was explained in a basic model. The primitive zero-compliance is a series-connected two suspensions; one of them has negative stiffness whose absolute value is equal to that of the other suspension with positive stiffness. The connection point becomes the detection point in measurement systems. Practically, at least one of the suspensions must be an active element equipped with an actuator. The measurement systems were classified from three viewpoints related to such suspension. Three examples of measurement system with a possible configuration were presented. Then, the control methods of achieving zero-compliance were explained in a mathematical model of the first example that used double series magnetic suspension to achieve the zero-compliance characteristic. The measurement results were presented which demonstrate the high resolution of the developed system.

The measurement systems have a unique characteristic that the force is estimated from the deflection of the detection point like by a deflection method, while the position of the point of application of force is maintained at the original position (with no applied force) like by a null method. This characteristic is expected to lead to an innovation in the field of force measurement.

This chapter has focused on a single-axis force measurement. Several multi-axis force measurement systems have been also developed [18, 19]. In addition, a measurement device with a cantilever for detecting force has been developed [20]. This device targets scanning microscopes with a novel rigorous force detection mechanism with multi-degree-of-freedom-of-motion zero-compliance suspension [20]. The third example shown in Section 3 also targets such microscopes. The next step will be the development of a multi-degree-of-freedom-of-motion zero-compliance suspension based on the MEMS technology.

Acknowledgements

This work was financially supported in part by JSPS KAKENHI Grant Numbers: 25630073, 17H03188 and 21K18162.

Thanks

The authors thank Mr. Sekine, Mr. Takabayashi, and Mr. Monjiyama for their help in designing and fabricating the apparatuses shown in Section 3 and conducting experimental works shown in Section 4.

Author details

Takeshi Mizuno*, Masaya Takasaki and Yuji Ishino
Saitama University, Saitama, Japan

*Address all correspondence to: mizar@mech.saitama-u.ac.jp

References

[1] Stefanescu DM. Handbook of Force Transducers. Berlin Heidelberg: Springer; 2011. 1-48 p. DOI: 10.1007/978-3-642-18206-9

[2] Kochsiek M, Gläser M, editors. Comprehensive Mass Metrology. Berlin: Wiley-VCH; 2000. 218-311 p

[3] Bentley JP, editor. Principle of Measurement Systems. 3rd ed. Harlow: Longman Science & Technology; 1995. 196-201 p

[4] Ohba R, editor. Intelligent Sensor Technology. Chichester: John Wiley & Sons; 1992. 13 p

[5] Andraea J, Sawlab A. Time synchronized measurement of multi-bridge force transducers. Measurement. 2001;**29**:105-111. DOI: 10.1016/S0263-2241(00)00031-2

[6] Hayashi T, Katase Y, Ueda K, Hoshino T, Suzawa H, Kobayashi M. Evaluation of tuning fork type force transducer for use as a transfer standard. Measurement. 2008;**41**:941-949. DOI: 10.1016/j.measurement.2008.01.002

[7] Kuma H, Sharma C, Arora PK, Moona G, Kumar A. Development and metrological characterization of a precision force transducer for static force measurement related to applications. Measurement. 2016;**88**: 76-77. DOI: 10.1016/j.measurement.2016.03.027

[8] Shinners SM. Modern Control System Theory and Design. New York: John Wiley & Sons, Inc.; 1992. 391-486 p

[9] Grantham WJ, Vincent TL. Modern Control Systems Analysis and Design. New York: John Wiley & Sons, Inc.; 1993. 175-222 p

[10] Wilkie J, Johnson M, Katebi R. Control Engineering. New York: Palgrave; 2002. 196-218, 277-307, 383-501 p

[11] Mizuno T, Hayashi Y, Ishino Y, Takasaki M. Proposal of force measurement using a zero-compliance mechanism. In: Proceedings of the XXI IMEKO World Congress; 30 August - 4 September 2015; Prague. Prague: Czech Technical University; 2015. pp. 317-320

[12] Mizuno T, Yoshitomi R. Vibration isolation system using zero-power magnetic suspension (1st report, principles and basic experiments). Transactions of the Japan Society of Mechanical Engineers (JSME) Series C (in Japanese). 2002;**68**:2599-260.4

[13] Mizuno T. Force measurement using zero-compliance mechanism. In: Proceedings of Dynamic and Design Conference; 8–26 August 2016, Ube. Tokyo: JSME; 2016. p. KL2. DOI: 10.1299/jsmedmc.2016.KL2

[14] Mizuno T, Ishino Y, Kawada H, Hayashi Y, Takasaki M, Hara M, et al. Force measurement using zero-compliance mechanism. SICE Journal of Control, Measurement, and System Integration. 2018;**11**(4):341-348. DOI: 10.9746/jcmsi.11.341

[15] Mizuno T, Sekine D, Takasaki M, Ishino Y. Force measurement using double series magnetic suspension. Transactions of the JSME (in Japanese). 2014;**80**(814):DR0163

[16] Mizuno T, Takasaki M, Ishino Y. Development of one-piece

zero-compliance mechanism for force measurement. In: Proceedings of the 20th IFAC World Congress; 9–14 July 2017, Toulouse. Amsterdam: Elsevier Ltd (IFAC PaperOnLine). Vol. 50(1). 2017. pp. 6067-6072. DOI: 10.1016/j.ifacol.2017.08.1380

[17] Monjiyama R, Mizuno T, Ishino Y, Takasaki M. Development of MEMS-based micro-force measurement system with zero-compliance mechanism. Transactions of the JSME (in Japanese). 2021;**87**:901. DOI: 10.1299/transjsme.21-00122

[18] Mizuno T, Takasaki M, Ishino Y. Fabrication of a three-dimensional force measurement system using double series magnetic suspension. In: Proceedings of the 7th IFAC Symposium on Mechatronic Systems; 6–8 September 2016, Loughborough. Amsterdam: Elsevier Ltd (IFAC PaperOnLine). Vol. 49(21). 2016. pp. 536-540. DOI: 10.1016/j.ifacol.2016.10.657

[19] Nahiyan MHA, Mizuno T, Takasaki M, Ishino Y, Hara M, Yamaguchi D. Modeling and validation of vertical direction force estimation with a three-dimensional force measurement instrument based on a zero-compliance mechanism. Sensors. 2019;**19**:799. DOI: 10.3390/s19040799

[20] Mizuno T, Hayashi Y, Takasaki M, Ishino Y, Yamaguchi D. Development of three-degree-of-freedom zero-compliance mechanism for micro force measurement with a cantilever. In: Proceedings of the 21st IFAC World Congress; 12–17 July 2020, Berlin (Virtual). Amsterdam: Elsevier Ltd (IFAC PaperOnLine). Vol. 53(2). 2020. pp. 8339-8344. DOI: 10.1016/j.ifacol.2020.12.1618

Chapter 5

Effect of Process Parameter Variations on Triangular Microcavity Fabrication Using Micro-EDM

Suresh Pratap and Somak Datta

Abstract

This research study examines the effects of process parameters on fabricated triangular micro holes using micro-EDM (Micro Electrical Discharge Machining). The micro-EDM process utilized a 400 μm tungsten carbide electrode shaped as an equilateral triangle. The machining parameter assessment involved measuring errors in the corner radius and the included angle of the microcavity. The corner radius of the triangular microcavity ranged from 5 to 15 μm. Simultaneously, the included angle varied between 61 and 64°. These variations were attributed to adjustments in capacitance, voltage, and feed settings, which spanned from 1 nF, 90 V, and 5 μm to 100 nF, 130 V, and 15 μm, respectively. Statistical analysis using ANOVA revealed that voltage and capacitance significantly impacted the alterations in corner radii and included the angle of the manufactured triangular microcavities. FESEM is used to analyze the recast layer formation on the edges of the microcavity. It will help to decrease the error in fabricating triangular cavities using micro EDM.

Keywords: advanced manufacturing process, micromachining, micro EDM, triangular microcavity, ANOVA

1. Introduction

Precision is essential in the field of micromanufacturing. High dimensional accuracy is an essential target requiring extreme attention to every aspect of production. Fabricating micro polygonal profiles is difficult to create using traditional machining processes [1]. Maintaining corner and angular precision in polygonal profile products may be difficult despite its importance. As a result, many manufacturers are looking at new technologies and techniques that will enable them to achieve more precision and accuracy in their micromanufacturing processes. Sophisticated machining processes can meet the demands of modern industry. Electrochemical machining

(ECM) [2], a technique that dissolves metal using a conductive fluid, is often used in the aerospace and medical sectors. Laser beam machining (LBM) is utilized in the electronics and automotive sectors to remove material with a strong laser. Ultrasonic machining (USM) frequently removes material using ultrasonic vibrations and produces micro-components. Electron beam machining is predominantly used in aerospace to remove material using an electron beam. For high-precision and high-speed machining in aerospace and medical applications, electrochemical discharge machining (ECDM) is a hybrid technique that combines electrical discharge machining (EDM) and electrochemical discharge machining (ECM). Electrical discharge machining (EDM) [3] is a non-contact technique that erodes material using electrical sparks. It is best suited for machining complicated structures and harder materials like titanium, carbide, and tool steels. Among these, EDM is particularly noteworthy for its ability to machine any conductive material, regardless of hardness, and create complex shapes with high precision. Micro-EDM is the most commonly used variant, especially for micro-structures (holes, slots) and complicated 3D profiles [4].

The Micro EDM process has two major limitations: its material removal rate is comparatively low compared to other processes, and the tool wear rate during EDM is high. Various approaches and methods are used for predicting and optimizing the TWR and MRR of micro EDM. Bellotti et al. [5] used data-driven methods to create regression models that could predict tool wear and material removal rates (MRR and TWR) in micro-EDM blind holes. By using process monitoring data as input for their regression models, the researchers reduced errors in predicting MRR and TWR by approximately 65 and 85%, respectively. Pragadish et al. [6] greatest impact on material removal rate (MRR) and tool wear rate (TWR) while drilling silicon steel and optimizing the micro EDM process parameters to achieve high MRR with reduced tool wear rate in coated tools and found dielectric cardanol oil has major influence and at gap voltage of 50 V, Green dielectric (%) of 05 and coating thickness of 1 μm, got optimum MRR of 9.69 mm3/min and a tool wear rate (TWR) of 1.09 mm3/min. Arunnath et al. [7] fabricated a hole on T6 aluminum alloy using aluminum 7075 nano boron carbide metal matrix composites using EDM, and ANOVA was used to analyze the experimental results to know the percentage of contribution of each parameter on MRR and TWR. Sanghani et al. [8] developed a mathematical model using a regression equation of MRR and TWR in EDM based on a fraction-of-energy approach. Vidya et al. [9, 10] investigated the dimensional accuracy of micro holes and micro-channels fabricated EN-24 alloy steel using die sinking EDM process and found that the micro holes have a roundness error of 46.33 μm, and the microchannels have a straightness tolerance of 13.51 μm. Geometric tolerance greatly affects the performance and lifetime of the mechanical parts. Microcavity was created by Abhinav et al. [11] discovered that the diametral length of the microcavity varied between 748 and 800 μm.

The manufactured micro holes using micro EDM show extremely less circularity variation. Near the edges of the machined holes, some recast layer development that varies between 6.5 and 7.5 μm was observed. According to Huan et al. [12], it is necessary to compensate for micro-electrode wear to maintain the dimensional consistency and accuracy of micro-hole arrays in micro-EDM drilling. Mouralova et al. [8] studied the fabrication of a precise slot of size 5000 × 170 μm in a copper foil with a thickness of 125 μm and used the same copper foil as a tool in micro EDM. Unune et al. [13] studied the dimensional accuracy and surface quality of micro-channels with low-frequency vibration assistance in micro-electro-discharge milling. They found that the discharge energy significantly affected the amount and size of the globules formed during μ-ED milling. Rafaqat et al. [14] produced non-circular holes of three shapes (square, triangle, and hexagon) in AISI D2 steel. They evaluated the performance of three response characteristics: material removal rate, tool wear rate, and land wear.

These limitations impact the dimensional accuracy of the machined feature. Researchers continuously improve the material removal rate and reduce the tool wear rate. In this investigation, the performance of micro EDM was evaluated in fabricating a polygonal structure, specifically a triangular profile. Non-circular shaped holes, commonly used in the mold industry, were selected as the machining profiles. The performance was evaluated against two triangle features, included angle between sides and corner radius, by varying the input parameters such as voltage, capacitance, and feed.

2. Experimental details

This study experimented on a MIKROTOOLS DT110 (hybrid micro-EDM), as shown in **Figure 1**, which includes micro-drilling, micro-milling, micro-grinding, micro-turning, and micro-EDM. The experiment used the Taguchi design, a robust design modeling method. Taguchi's philosophy of robust design focuses on making products and processes less sensitive to sources of variation, such as environmental conditions or manufacturing fluctuations. An L8 orthogonal array was constructed for this experiment. Orthogonal arrays are pre-arranged tables that allow researchers to explore multiple factors and their interactions with relatively few experimental runs, thereby saving time and resources. The equilateral triangle tool of tungsten carbide having a side of 400 μm and a corner radius of 8 μm, as shown in **Figure 2(a)**, is used. The designed experiment was further optimized using ANOVA, a statistical method used to analyze the variance in data by partitioning the total variability into

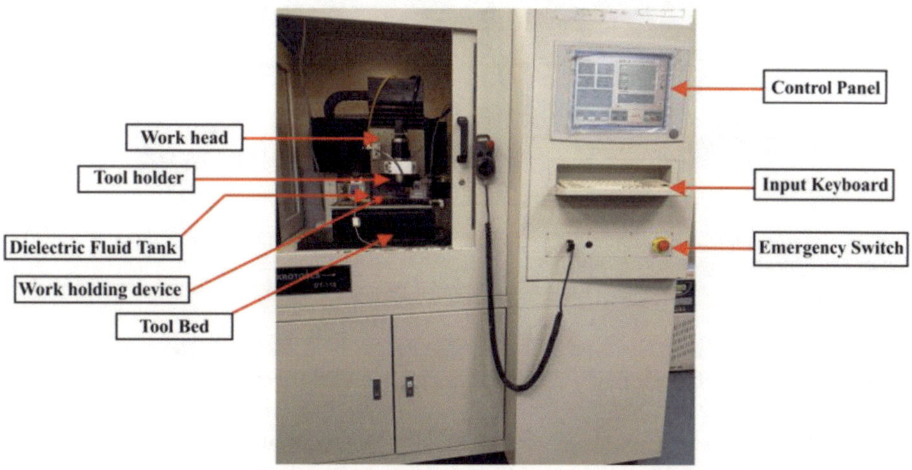

Figure 1.
Hybrid micromachining Centre-MIKROTOOLS DT110.

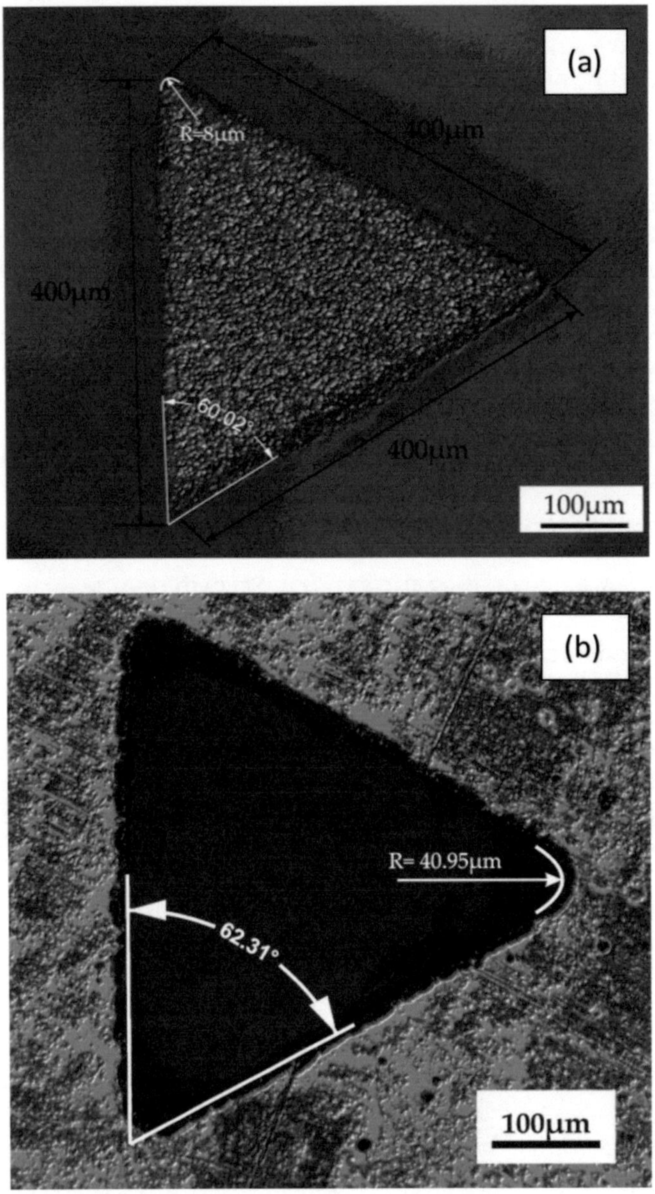

Figure 2.
(a) Tool of an equilateral triangle of side 400 μm, included angle 60°, and initial corner radius of 8 μm (b) triangular hole generated through micro-EDM.

different sources, such as the variability between groups and within groups. In optimization, ANOVA helps identify significant factors and interactions that influence the response variable, enabling practitioners to make informed decisions to improve processes and achieve optimal outcomes [15, 16]. The corner radius and included side angle of the triangular hole are calculated using image analysis with the help of the Zeiss AXIO Scope A1 optical microscope equipped with the inbuilt AXIO Vision software (**Table 1**).

Sl. No	Voltage (V)	Capacitance (nF)	Feed (μm/sec.)	Corner radius (μm)	Included angle (°)
1	90	10	5	35.08	61.12
2	90	100	5	37.45	62.31
3	90	10	20	35.24	61.27
4	90	100	20	38.20	62.78
5	120	10	20	39.71	61.78
6	120	100	20	40.05	63.69
7	120	10	5	40.95	61.41
8	120	100	5	41.08	63.59

Table 1.
L8 Taguchi design of experiment with responses corner radius and included angle.

Source	DF	Adj SS	Adj MS	F-Value	P-Value	Contribution %
Voltage (V)	1	3.1001	3.1001	25.68	0.007	4.937
Capacitance (nF)	1	59.187	59.187	490.32	0	94.259
Feed (μm/sec.)	1	0.0221	0.0221	0.18	0.691	0.035
Error	4	0.4829	0.1207			0.769
Total	7	62.792				

Table 2.
Analysis of variance for corner radius of triangular microcavity.

3. Results and discussions

3.1 Corner radius

In the field of micro-EDM, the corner radius of a triangle microcavity is of utmost importance. For numerous reasons, it is crucial to precisely manage this dimension, defined as the curvature or rounding at the microcavity's corners. A lower radius produces sharper corners, which has an impact on the geometric correctness of the workpiece. It has an impact on machining accuracy, first and foremost. **Table 2** defines that in the case of corner radius, voltage contributed 4.9%, capacitance 94%, and feed 0.03%.

Figure 3 presents a main effect plot that graphically illustrates the impact of varying key process parameters, namely input voltage (V), capacitance (μF), and feed rate (μm/sec), on the corner radius of triangular microcavities manufactured using Micro-EDM. The plot provides a clear depiction of how changes in these parameters influence the corner radius of the microcavities. The plot represents two different input voltage levels, specifically 90 and 120 V. Each voltage level is associated with a data point on the graph. It is evident that as the input voltage increases from 90 to 120 V, there is a noticeable upward trend in the corner radius of the microcavities. It suggests that higher input voltage settings result in larger corner radii. As capacitance increases from 10 to 100 μF, the corner radius of the microcavities also increases.

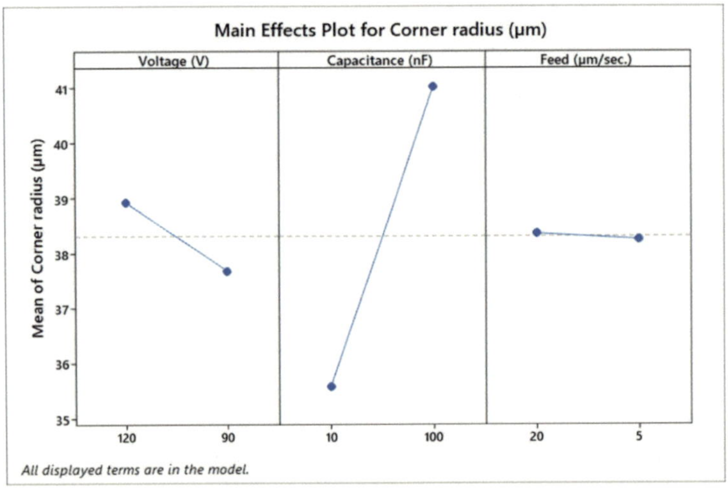

Figure 3.
Main effect plot for corner radius.

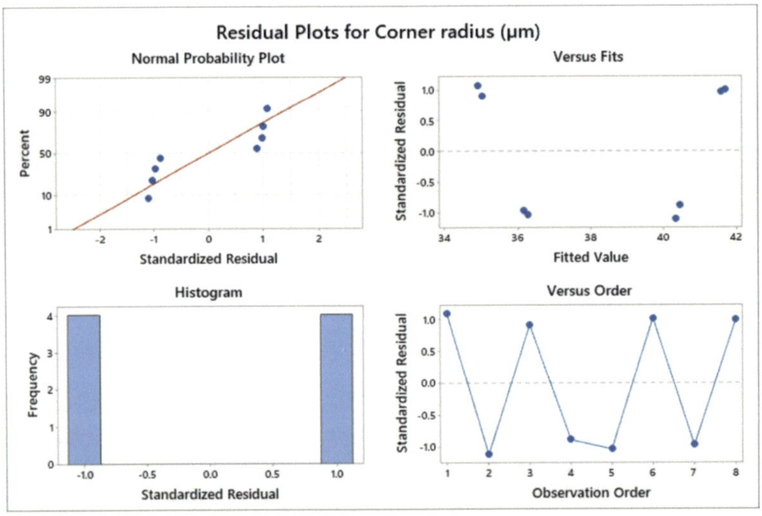

Figure 4.
Residual plot for corner radius.

An upward slope on the graph represents this relationship. It indicates that higher capacitance settings lead to larger corner radii.

The effect of feed rate on the corner radius appears to be relatively minimal compared to the other parameters. There is only a slight change in the corner radius as the feed rate varies between 5 and 15 μm/sec. It is represented by a nearly horizontal line on the graph, suggesting that changes in feed rate have less pronounced effects on the corner radius when compared to input voltage and capacitance variations. **Figure 4** provides a comprehensive view of the residuals from the model used to predict the corner radius of triangular microcavities. It includes visualizations and analyses to assess whether the residuals meet key assumptions, such as normality and independence, which are crucial for the reliability of the model's predictions.

These diagnostic tools are essential for validating the model and ensuring its suitability for practical applications in micro-EDM microcavity fabrication. **Figure 5** shows an interaction plot for corner radius, which clearly shows the interrelation of voltage, feed, and capacitance with the combination of one to another in a single figure.

3.2 Included angle

The included angle of a triangular microcavity is a critical parameter of interest. This angle, formed by the two sidewalls of the triangular cavity, holds immense importance in micro EDM for several reasons. Firstly, it directly affects the shape and dimensions of the machined feature, which is pivotal in achieving precise geometries in micro components. Secondly, the included angle impacts the EDM process stability and efficiency, influencing factors such as discharge energy distribution and material removal rates. Also, studying the included angle is essential for optimizing electrode design and machining parameters, ensuring that the microcavities meet various industries' specific requirements, including microelectronics, medical device manufacturing, and microfluidics. **Table 3** defines that in the case of the included angle of the triangular microcavity, voltage contributed 24.3%, capacitance 61%, and feed 0.5%.

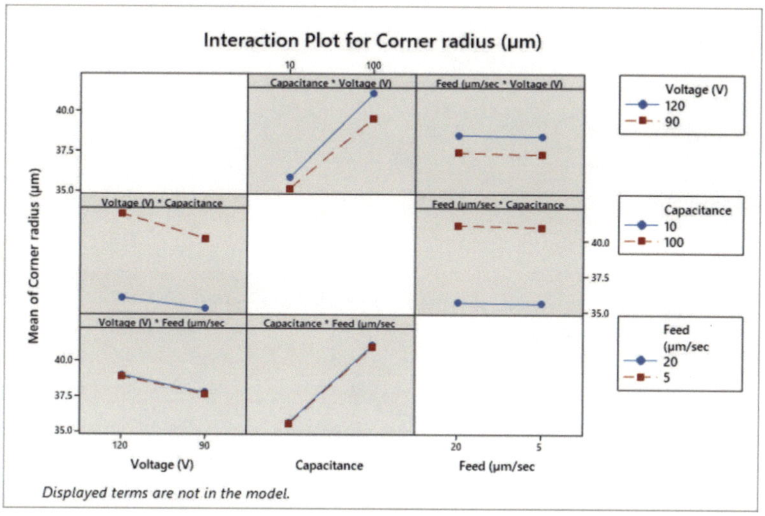

Figure 5.
Interaction plot for corner radius.

Source	DF	Adj SS	Adj MS	F-Value	P-Value	Contribution %
Voltage (V)	1	5.445	5.445	7.25	0.055	24.365
Capacitance (nF)	1	13.7812	13.7812	18.34	0.013	61.668
Feed (μm/sec.)	1	0.1152	0.1152	0.15	0.715	0.515
Error	4	3.0061	0.7515			13.452
Total	7	22.3475				

Table 3.
Analysis of variance for the included angle of triangular microcavity.

Figure 6 illustrates the impact of varying input voltage (V), capacitance (μF), and feed rate (μm/sec) on the included angle of triangular microcavities fabricated using micro-EDM. Higher input voltage and capacitance levels result in larger corner radii, as evidenced by the upward trends in the graph. In contrast, changes in feed rate between 5 and 20 μm/sec have a comparatively minimal effect on the included angle, as indicated by the nearly horizontal line on the plot. **Figure 7** describes the residual plot for the included angle of microcavity. **Figure 8** shows an interaction plot for the

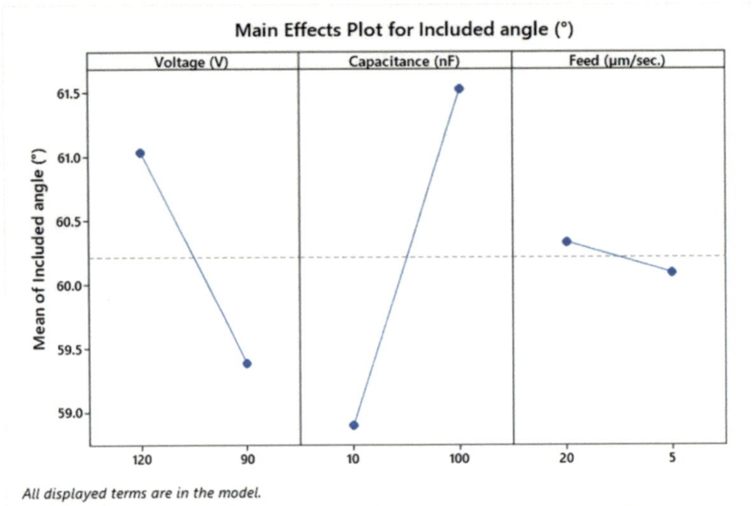

Figure 6.
Main effect plot for included angle.

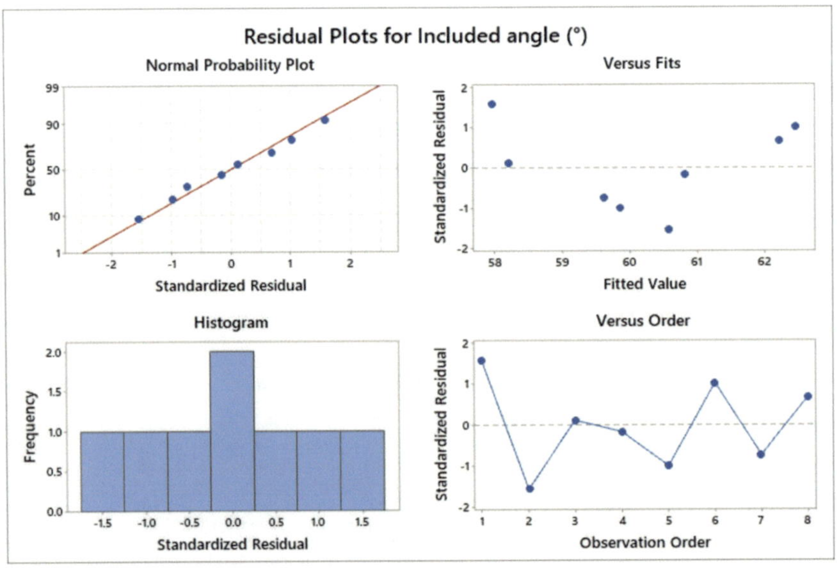

Figure 7.
Residual plot for included angle.

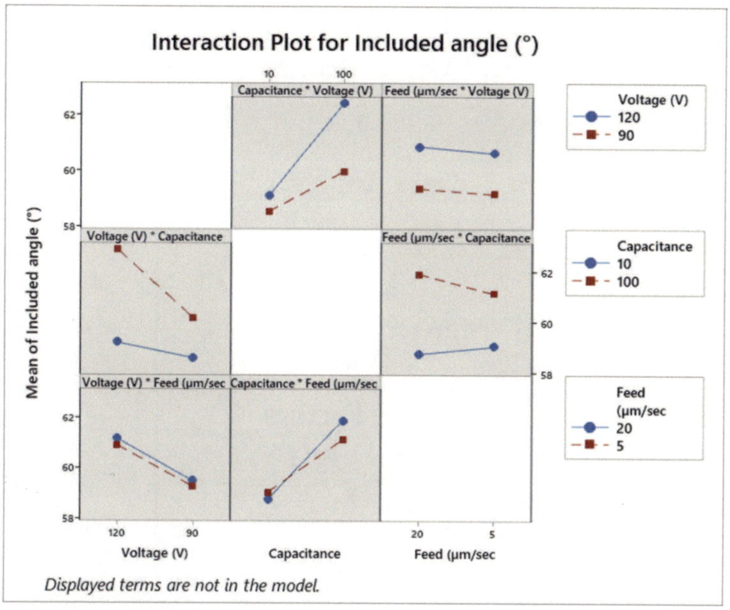

Figure 8.
Interaction plot for included angle.

included angle of microcavity, which clearly shows the interrelation of voltage, feed, and capacitance with the combination of one to another in a single figure.

3.3 Optimization using ANOVA

Optimization using Analysis of Variance (ANOVA) in Minitab involves using statistical techniques to determine the optimal settings or conditions for a process or system by analyzing different factors or variables that affect the response variable—the model to perform optimization using ANOVA, as shown in **Figure 9**. The optimum solution for both corner radius and included angle of triangular microcavity achieved through ANOVA is 90 V, 10 nF, and 5 µm/sec, as in **Figure 10**.

4. Conclusions

This investigation has systematically probed the influence of input parameters on the corner radius and included angle of triangular micro cavities produced via micro-electrical discharge machining (micro EDM). The empirical observations offer the following conclusions.

- A discernible relationship between the corner radius of the microcavity and the included angle was discerned as the corner radius expanded from 5 to 15 µm, resulting in an associated variation in the included angle from 62 to 67°. The dynamic response elucidates the pronounced sensitivity of the micro EDM process to modifications in input parameters, thus accentuating the paramount importance of precise parameter manipulation to attain the desired geometrical outcomes.

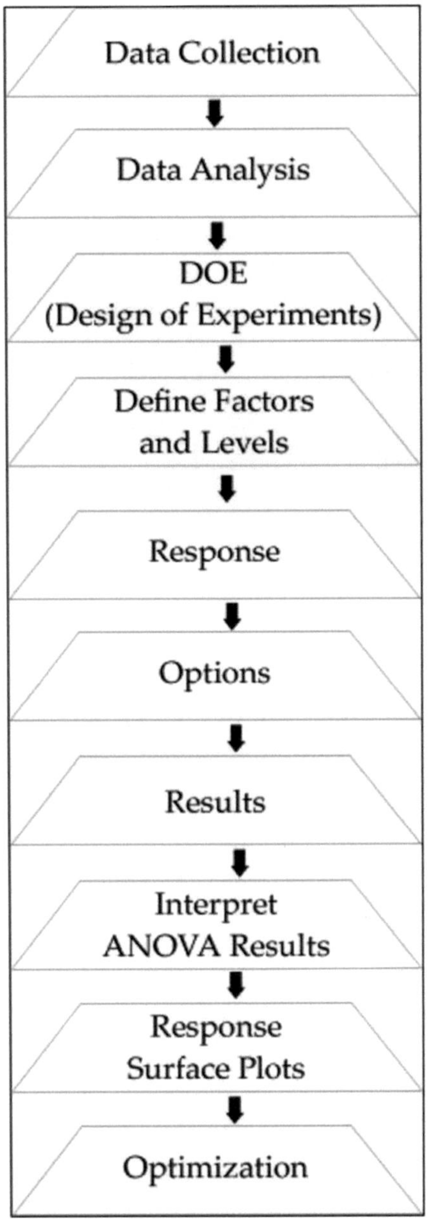

Figure 9.
Process optimization using ANOVA.

- Applying statistical and analysis of variance (ANOVA) techniques underscored that capacitance substantially influenced both the corner radius and included angle.

- The investigation unveiled that the most favorable operational conditions for minimizing variation in microcavity geometry comprised a voltage of 90 V, a capacitance of 10 nF, and a feed rate of 5 μm/sec. is preferable to include a Conclusion(s) section which will summarize the content of the book chapter.

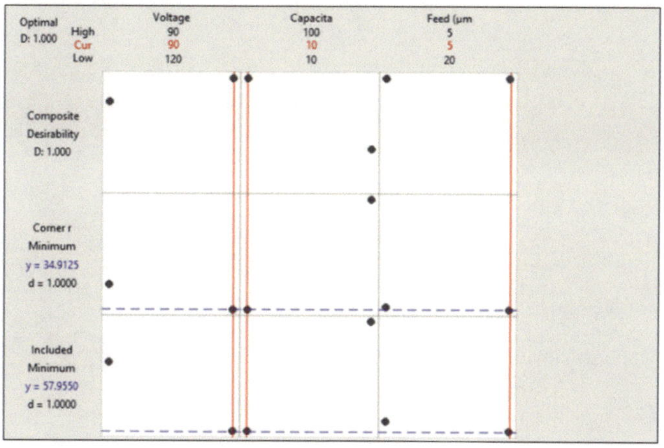

Figure 10.
Optimal solution.

Acknowledgements

All authors fulfill the criteria for authorship.

Funding

There is no funding received.

Conflict of interest

The authors declare no conflict of interest.

Notes/thanks/other declarations

Special thanks to Department of Production and Industrial Engineering, Birla Institute of technology Mesra for providing the research facilities.

Author details

Suresh Pratap* and Somak Datta
Birla Institute of Technology Mesra, Ranchi, India

*Address all correspondence to: sureshpratap@yahoo.com

References

[1] Kumar D, Singh NK, Bajpai V. Recent trends, opportunities and other aspects of micro-EDM for advanced manufacturing: A comprehensive review. Journal of the Brazilian Society of Mechanical Sciences and Engineering. 2020;**42**(5):222

[2] Kim BH, Park BJ, Chu CN. Fabrication of multiple electrodes by reverse EDM and their application in micro ECM. Journal of Micromechanics and Microengineering. 2006;**16**(4):843-850

[3] Mastud SA. Reverse micro-EDM. In: Kibria G, Jahan MP, Bhattacharyya B, editors. Micro-Electrical Discharge Machining Processes. Singapore: Springer Singapore; 2019. pp. 93-124 (Materials Forming, Machining and Tribology). Available from: http://link.springer.com/10.1007/978-981-13-3074-2_5

[4] Liu K, Lauwers B, Reynaerts D. Process capabilities of micro-EDM and its applications. The International Journal of Advanced Manufacturing Technology. 2010;**47**:11-19

[5] Bellotti M, Wu M, Qian J, Reynaerts D. Tool wear and material removal predictions in micro-EDM drilling: Advantages of data-driven approaches. Applied Sciences. 2020;**10**(18):6357

[6] Pragadish N, Kaliappan S, Subramanian M, Natrayan L, Satish Prakash K, Subbiah R, et al. Optimization of cardanol oil dielectric-activated EDM process parameters in machining of silicon steel. Biomass Conversion and Biorefinery. 2022;**13**:1-10

[7] Arunnath A, Madhu S, Tufa M. Experimental investigation and optimization of material removal rate and tool Wear in the machining of aluminum-boron carbide (Al-B4C) nanocomposite using EDM process. Advances in Materials Science and Engineering. 2022;**2022**:1-11

[8] Sanghani C, Acharya G, Kothari K. Modelling of material removal rate and tool wear rate in EDM based on a fraction of energy approach. Advances in Materials and Processing Technologies. 2022;**8**(1):130-146

[9] Vidya S, Wattal R, Venkateswara RP. Software based analysis of angularity error for polygonal micro-cavities produced by die-sinking EDM. Materials Today: Proceedings. 2022;**56**:559-562

[10] Vidya S, Wattal R, Venkateswara RP. Evaluation of material removal rate for different cavity geometries produced by die-sinking electrical discharge machining. Materials Today: Proceedings. 2023;**72**:1132-1135

[11] Kumar A, Kumar D, Singh NK. Fabricating micro-holes through micro-EDM process and their circularity testing. In: Recent Trends in Product Design and Intelligent Manufacturing Systems: Select Proceedings of IPDIMS 2021. Springer; 2022. pp. 731-737

[12] Huan L, Jicheng B, Yan C, Guozheng Z, Shaojie H. Micro-electrode wear and compensation to ensure the dimensional consistency accuracy of micro-hole array in micro-EDM drilling. The International Journal of Advanced Manufacturing Technology. 2020;**111**:2653-2665

[13] Unune DR, Nirala CK, Mali HS. Accuracy and quality of micro-holes in vibration assisted

micro-electro-discharge drilling of Inconel 718. Measurement. 2019;**135**:424-437

[14] Rafaqat M, Mufti NA, Saleem MQ, Ahmed N, Hussain A. Electric discharge machining of non-circular through-holes: Material removal and tool wear analysis. Journal of the Brazilian Society of Mechanical Sciences and Engineering. 2023;**45**(3):135

[15] Pratap S, Datta S. Fabrication of array microelectrodes: Achieving geometrical characteristics using reverse micro-EDM process. Physica Scripta. 2023;**98**(8):085003

[16] Pratap S, Datta S, Sharma V. Determination of surface roughness of micro rods fabricated by reverse μEDM using dimensional analysis. The International Journal of Advanced Manufacturing Technology. 2023;**128**(7):3649-3659